Securing Industrial Control Systems and Safety Instrumented Systems

A practical guide for safeguarding mission and safety critical systems

Jalal Bouhdada

Securing Industrial Control Systems and Safety Instrumented Systems

Group Product Manager: Pavan Ramchandani
Publishing Product Manager: Neha Sharma
Book Project Manager: Ashwini C
Senior Editors: Romy Dias and Apramit Bhattacharya
Technical Editor: Arjun Varma
Copy Editor: Safis Editing
Proofreader: Apramit Bhattacharya
Indexer: Pratik Shirodkar
Production Designer: Jyoti Kadam
DevRel Marketing Coordinator: Marylou De Mello

First published: August 2024

Production reference: 1310724

Published by Packt Publishing Ltd.
Grosvenor House
11 St Paul's Square
Birmingham
B3 1RB, UK

ISBN 978-1-80107-881-8

www.packtpub.com

To my cherished wife, my dear daughters, my beloved parents, and my brothers.

Thank you for the memories that brighten my days, for the unconditional love that sustains my soul, and for the unwavering support that propels me forward. Each page of this book is imbued with the gratitude I hold for you all, as my life's most precious guides and companions.

Additionally, I would like to extend my gratitude to the ICS cybersecurity community for their invaluable guidance and continuous support during the research and writing of this book.

Special thanks to the book's technical reviewers and to the Packt project team for their flexibility and dedication.

Foreword

It's a true honor to introduce this insightful book on practical **Safety Instrumented Systems (SISs)** cybersecurity, authored by a colleague and friend I've had the pleasure of knowing for many years. Our shared background in systems design, asset management, and security within the energy, chemical, and critical infrastructure sectors has given me a firsthand look at his exceptional expertise and commitment to securing **Industrial Control Systems (ICSs)**. We both worked earlier in our careers at major Dutch companies that operate globally, which brought us together early on in technical discussions specific to the process control and safety domain of plant operations.

The author is a highly skilled expert in cybersecurity for process control domains and networks. His technical knowledge is extensive, and his warm-hearted nature and kindness have made him a trusted advisor to many. His genuine interest in listening to and addressing concerns has earned him respect and admiration across the industry.

Throughout his career, the author has provided invaluable support to companies in the oil and gas, chemicals, maritime, power and utilities sectors, and more. His ability to simplify complex cybersecurity concepts and present them in a practical, understandable way has empowered organizations to better protect their critical systems and assets. He has a remarkable talent for discerning what truly matters in control systems and cybersecurity, separating the essential from the subjective, and conveying these insights with clarity.

This book is a testament to the author's extensive knowledge and experience. It offers a comprehensive guide to the intricacies of securing SISs within industrial environments, a subject of top importance in today's increasingly interconnected world. His deep understanding of both the technical and operational aspects of SIS cybersecurity ensures that this book is not only informative but also highly practical, providing readers with actionable insights and strategies to enhance the security of their systems.

What sets this book apart is the author's unique ability to bridge the gap between theory and practice. His firsthand experience as an asset owner-operator, combined with his extensive work helping various companies worldwide, has provided him with a wealth of practical knowledge and real-world insights. These experiences are covered throughout the book, offering readers a rich tapestry of case studies, best practices, and lessons learned from the front lines of industrial cybersecurity.

In an era where cybersecurity threats are constantly evolving and becoming increasingly sophisticated, the importance of securing SISs cannot be overstated. These systems play a crucial role in ensuring the safety of industrial processes, and any compromise to their integrity can have catastrophic consequences. The author's approach to SIS cybersecurity, rooted in years of experience assessing and securing different vendor solutions, makes this book a solid resource for anyone involved in the design, implementation, or management of these critical systems.

As someone who has had the privilege of collaborating with the author and witnessing firsthand his dedication, expertise, and passion for cybersecurity, I can attest to his qualifications and the value of the insights presented in this book. His contributions to the field are widely recognized and respected within the industry, and this book is yet another testament to his commitment to advancing the state of SIS cybersecurity.

In closing, I highly recommend this book and encourage you to explore its pages with confidence. Whether you are a seasoned professional or new to the world of SIS cybersecurity, you will find this book to be an indispensable guide, filled with practical advice, expert insights, and the wisdom of someone who has dedicated his career to making our industrial environments safer and more secure.

Sincerely,

Marco (Marc) Ayala

Contributors

About the author

Jalal Bouhdada is a renowned international expert in the realm of **Industrial Control Systems (ICSs)** cybersecurity, with a deep-rooted passion for cybersecurity, cutting-edge technology, and the dynamic world of startups. As a founder, investor, and board advisor, Jalal is dedicated to driving significant advancements in cybersecurity. He is committed to fostering innovation within the industry and aiding critical infrastructure organizations in securing their digital landscapes.

Since founding Applied Risk in 2012, Jalal has been at the forefront of the company, steering its strategy toward becoming a leader in industrial security services and innovative product development. Under his leadership, Applied Risk has tackled numerous intricate ICS cybersecurity projects for prominent global clients, including some of the largest names in the industrial and utilities sectors.

Recognized as a global thought leader in industrial control systems security and critical infrastructure protection, Jalal actively contributes to the cybersecurity community. He is a member of several prestigious security societies and has co-authored pivotal ICS security best practice guidelines for notable organizations such as ENISA, ISA, and the **European Energy ISAC (EE-ISAC)**. Additionally, Jalal is a sought-after speaker who frequently shares his expertise with both private and public audiences worldwide, inspiring a new generation of cybersecurity professionals.

About the reviewers

Denrich Sananda is an instrumentation engineer with a distinguished qualification from Harvard Business School and boasts an impressive 24-year career dedicated to process safety automation. Renowned for his leadership prowess, Denrich has spearheaded high-potential teams within complex business environments, safeguarding high-value assets across sectors including oil and gas, utilities, fertilizers, petrochemicals, and refineries.

With a focus on risk mitigation and management, Denrich specializes in providing comprehensive solutions for functional safety and **Operational Technology (OT)** security within ICS environments. Leveraging industry standards such as ISA/IEC 61511, ISA/IEC 62443, NIST, and NERC CIP, he ensures the robust protection of critical infrastructure.

Denrich's hands-on experience encompasses the execution, commissioning, and troubleshooting of numerous safety system projects, underscoring his practical expertise in the field.

As an accredited TÜV Rheinland program trainer for SIS, Denrich is committed to sharing his knowledge and expertise. He has conducted numerous functional safety seminars and training sessions aimed at fostering awareness of the IEC 61508 and IEC 61511 lifecycle approaches.

Passionate about safety and security, Denrich is a sought-after speaker at conferences and forums, where he actively engages in discussions surrounding functional safety and OT/ICS cybersecurity, further solidifying his reputation as a thought leader in the field.

Marc Ayala is a process automation professional with over 25 years of experience working in petrochemical facilities where he designed, engineered, and maintained process automation, safety systems, and integrated networks. Marco is active in the oil and gas sector, chemicals industry, and maritime domain including offshore facilities. Mr. Ayala is an established and respected instructor for ISA cyber courses. He is a member of and contributor to the AMSC Cybersecurity efforts as Chair, an InfraGard member, and sector chief of the Maritime Domain Cross-Sector Council for ports and terminals.

Paul Smith has spent close to 20 years in the automation control space, tackling the "red herring" problems that were thrown his way. Unique issues such as measurement imbalances resulting from flare sensor saturation, database migration mishaps, EEPROM production line failures, and many more. This ultimately led to the later part of his career where he has spent most of his time in the industrial cybersecurity space pioneering the use of new security technology in the energy, utility, and critical infrastructure sectors, and helping develop cybersecurity strategies for some of the world's largest industrial organizations and municipalities.

Ron Brash is a household name when it comes to ICS/OT cybersecurity and embedded vulnerability research. He was instrumental in creating the datasets for the S4 ICS Detection Challenges, received the Top 40 under 40 award for Engineering Leaders 2020 from Plant Engineering, was an embedded developer at Tofino Security, advised several large asset owners in a variety of industries on OT security, and brought a number of products to market.

With 45 years of experience in process automation, **Sinclair Koelemij** brings a wealth of expertise, with 25 years dedicated to process control and an additional 20 years specializing in networking, security, and risk management for process automation systems. During his extensive 43-year tenure at Honeywell, he played key roles in servicing, engineering, and securing diverse control and process safety solutions, spanning basic and advanced control systems. Sinclair also possesses hands-on experience in software development and the implementation of control and safety solutions for over 100 installations owned by various asset owners, varying in scale from smaller setups with fewer than 1,000 I/O points to large installations exceeding 100,000 I/O points. Furthermore, Sinclair holds several patents in the field of cyber-physical risk evaluation and mitigation.

Younes Dragoni is a highly skilled and experienced professional specializing in ICS cybersecurity with a passion for supporting the cybersecurity maturity journey of major critical infrastructures worldwide.

Having started his career as a security researcher, Younes has an impressive track record of over 20 advisory publications, working with renowned companies such as Mitsubishi Electric, Siemens, Rockwell Automation, Emerson, GE, and Philips Healthcare. His deep technical knowledge and research contributions have made him a trusted authority in the field.

Making a transition into the business development field, Younes has been instrumental in driving growth, both in terms of revenue and strategic partnerships. He understands the unique challenges faced by organizations operating critical infrastructures and is committed to delivering innovative solutions that address their cybersecurity needs.

Beyond his professional accomplishments, Younes also serves as an advisor and mentor for early-stage startups and national institutions, leveraging his expertise to guide and shape their cybersecurity strategies. He is an active member of the World Economic Forum, contributing to discussions on the future of cybersecurity, and serves as a board member at the ISACA Swiss Chapter.

Table of Contents

Part 1: Safety Instrumented Systems

1

2

3

SIS Security Design and Architecture   51

Part 2: Attacking and Securing SISs

4

Hacking Safety Instrumented Systems   93

5

Securing Safety Instrumented Systems   131

Part 3: Risk Management and Compliance

6

Cybersecurity Risk Management of SISs   165

7

Security Standards and Certification   193

8

Preface

Industrial Control Systems (ICSs) form the backbone of modern industry, facilitating the automation, monitoring, and management of critical infrastructure sectors including oil and gas, chemicals, power generation, and manufacturing. Within this context, **Safety Instrumented Systems (SISs)** play a pivotal role in ensuring operational safety by preventing hazardous situations and reducing the likelihood of catastrophic failures. However, as the convergence of **Information Technology (IT)** and **Operational Technology (OT)** accelerates, the inherent vulnerabilities in ICS and SIS have exposed industries to a new dimension of cybersecurity threats.

The expanding digital footprint of industrial environments has brought remarkable gains in efficiency, productivity, and insight. Yet, it has also widened the attack surface, placing critical safety functions at risk. Cyberattacks on ICSs/SISs can result in devastating consequences, including process disruptions, equipment damage, environmental hazards, and even loss of life. Securing these systems requires a holistic understanding of both the technical and strategic aspects of cybersecurity, blending traditional safety measures with emerging security frameworks.

Securing Industrial Control Systems and Safety Instrumented Systems: A practical guide for safeguarding mission and safety critical systems is designed to offer an in-depth exploration of this challenging landscape. This book aims to provide practical guidance, strategic insights, and actionable steps to protect your ICS/SIS environment effectively.

The key objectives of this book are as follows:

- **Highlight the convergence**: Explore how the convergence of IT and OT creates both opportunities and challenges for securing industrial environments

- **Understand SIS fundamentals**: Provide a comprehensive overview of SISs, their architecture, and how they integrate with ICS

- **Explore cybersecurity risks and threats**: Identify the specific cybersecurity risks facing SISs within ICS environments, emphasizing the unique characteristics that make these systems vulnerable

- **Implement defense strategies**: Present practical strategies and solutions to secure SISs, leveraging best practices in cybersecurity and safety engineering

- **Build cyber resilience**: Advocate for a security resilience that blends safety practices with cybersecurity readiness, emphasizing the importance of people and processes alongside technology to be well equipped for emerging cyber threats

Who this book is for

This book is aimed at the following roles:

- **Industrial automation engineers**: To deepen their understanding of cybersecurity risks and cover how to integrate security into SIS design and operations

- **IT and OT security professionals**: To help them grasp the unique challenges of securing industrial environments and implement tailored cybersecurity strategies

- **Safety managers and process engineers**: To help them incorporate cybersecurity measures into existing safety frameworks and protocols

- **Policy makers and regulators**: To help them develop informed policies that ensure the resilience and security of critical infrastructure

What this book covers

Chapter 1, Introduction to Safety Instrumented Systems (SISs), lays the foundation by exploring what **Safety Instrumented Systems** (**SISs**) are and their crucial role in safeguarding industrial processes. We delve into the principles of functional safety, outlining key components, functions, and how SIS integrates with **Industrial Control Systems** (**ICSs**).

Chapter 2, SIS Evolution and Trends, traces the historical evolution of SIS, from early mechanical safeguards to modern electronic SIS. We also discuss emerging trends such as the convergence of IT and OT, the impact of new technologies, and the increasing adoption of integrated safety and control systems.

Chapter 3, SIS Security Design and Architecture, provides a comprehensive guide to designing and architecting SIS. Key topics include risk assessment, **Safety Integrity Levels** (**SILs**), redundancy models, and the integration of SIS with **Distributed Control Systems** (**DCSs**) and **Programmable Logic Controllers** (**PLCs**).

Chapter 4, Hacking Safety Instrumented Systems, is an eye-opening chapter in which we uncover the methods, tactics, and motivations of attackers targeting SISs. We analyze real-world case studies, explain common vulnerabilities, and discuss how cyberattacks can lead to catastrophic safety failures.

Chapter 5, Securing Safety Instrumented Systems, builds on the previous chapter and presents practical strategies and best practices to secure SISs. From network segmentation and anomaly detection to secure coding practices and incident response, this chapter offers a comprehensive approach to safeguarding critical safety systems.

Chapter 6, Cybersecurity Risk Management of SISs, delves into risk management, which is the cornerstone of effective SIS security. This chapter provides a systematic framework for identifying, assessing, and mitigating cybersecurity risks in SISs. We introduce risk assessment methodologies including HAZOP and LOPA and discuss how to prioritize controls based on their potential impact.

Chapter 7, Security Standards and Certification asserts, that compliance with standards and regulations is key to ensuring the safety and security of SIS. This chapter offers an overview of international standards including IEC 61511 and NIST 800-82, along with relevant legislation. We also provide practical guidance on implementing and adhering to these standards.

Chapter 8, The Future of ICS and SIS: Innovations and Challenges, is where we curate a comprehensive list of additional resources, including books, whitepapers, webinars, and industry organizations. These resources will help you deepen your understanding and stay updated on the latest trends in SIS security.

To get the most out of this book

This book assumes a foundational understanding of SISs and ICSs. Proficiency with Windows, macOS, or Linux operating systems will help you make the most of the practical insights provided.

Software/hardware covered in the book	Operating system requirements
MOSAIC M1S COM https://www.reersafety.com/en/software-2/	Windows, macOS, or Linux
Reer MOSAIC M1S Safety PLC https://www.reersafety.com/en/categories/safety-controllers-and-interfaces/	

Get in touch

Feedback from our readers is always welcome.

General feedback: If you have questions about any aspect of this book, email us at customercare@packtpub.com and mention the book title in the subject of your message.

Errata: Although we have taken every care to ensure the accuracy of our content, mistakes do happen. If you have found a mistake in this book, we would be grateful if you would report this to us. Please visit www.packtpub.com/support/errata and fill in the form.

Piracy: If you come across any illegal copies of our works in any form on the internet, we would be grateful if you would provide us with the location address or website name. Please contact us at copyright@packtpub.com with a link to the material.

If you are interested in becoming an author: If there is a topic that you have expertise in and you are interested in either writing or contributing to a book, please visit authors.packtpub.com.

Share Your Thoughts

Once you've read *Securing Industrial Control Systems and Safety Instrumented Systems*, we'd love to hear your thoughts! Scan the QR code below to go straight to the Amazon review page for this book and share your feedback.

https://packt.link/r/1-801-07881-5

Your review is important to us and the tech community and will help us make sure we're delivering excellent quality content.

Download a free PDF copy of this book

Thanks for purchasing this book!

Do you like to read on the go but are unable to carry your print books everywhere?

Is your eBook purchase not compatible with the device of your choice?

Don't worry, now with every Packt book you get a DRM-free PDF version of that book at no cost.

Read anywhere, any place, on any device. Search, copy, and paste code from your favorite technical books directly into your application.

The perks don't stop there, you can get exclusive access to discounts, newsletters, and great free content in your inbox daily

Follow these simple steps to get the benefits:

1. Scan the QR code or visit the link below

https://packt.link/free-ebook/978-1-80107-881-8

2. Submit your proof of purchase
3. That's it! We'll send your free PDF and other benefits to your email directly

Part 1:
Safety Instrumented Systems

This book begins with a comprehensive introduction to **Safety Instrumented Systems** (**SISs**), covering essential safety and cybersecurity concepts as well as terminology specific to process safety cybersecurity. The goal of the initial chapter is to establish a solid foundation of knowledge, enabling readers to delve deeper into more complex topics in subsequent chapters. *Chapter 2* addresses and clarifies common misconceptions about SIS cybersecurity to ensure a clear understanding before progressing to more detailed discussions. *Chapter 3* explores the security design and architecture, including protocols and best practices, emphasizing the secure-by-design principles.

This part has the following chapters:

- *Chapter 1, Introduction to Safety Instrumented Systems (SISs)*
- *Chapter 2, SIS Evolution and Trends*
- *Chapter 3, SIS Security Design and Architecture*

This structure ensures a progressive learning experience, equipping readers with both the theoretical and practical aspects of SIS cybersecurity.

1

Introduction to Safety Instrumented Systems (SISs)

Industrial control systems (ICSs) have become an increasingly pressing concern due to emerging cyber threats and the prevalence of legacy devices that lack the security to protect against modern threat vectors. Cyberattacks have struck assets of all sizes and verticals, bringing an end to the era of denial and myths about the security of industrial installations.

Safety instrumented systems (SISs) are considered the crown jewels and last layer of defense for many **Critical Infrastructures** (CIs) such as oil and gas, chemicals, power, manufacturing, and maritime to name a few.

For years, they have operated in isolation using technologies and protocols that were designed without security in mind and focusing primarily on operations conventional functional safety requirements that are not sufficient to protect against motivated, capable, and well-funded adversarial cyber threats.

Nowadays, modern process facilities are significantly interconnected due to the **Information Technology** (IT) and **Operational Technology** (OT) convergence, and the widespread adoption of **Internet Protocol** (IP) based technologies. Furthermore, access to vendor documentation and system specifications is no longer exclusive to a select group of asset owners, operators, and **Original Equipment Manufacturers** (OEMs). This renders an SIS increasingly vulnerable to cybersecurity attacks by adversaries seeking to manipulate or disrupt its operations.

The importance of cybersecurity for an SIS has only recently started to gain broader attention on C-suite agendas within organizations, primarily driven by the observation of a number of prominent cyber incidents and near-misses in recent years.

In this chapter, we're going to cover the following main topics together:

- Understanding SIS
- What is ICS cybersecurity?
- Exploring relevant cybersecurity and functional safety standards
- Examining the safety and cybersecurity lifecycle

Understanding SIS

The main goal of this chapter – and this book – is not to provide an extensive education on the engineering specifics of SISs, as many resources and publications already exist on this subject and have been available for some time. We will instead focus on what you need to understand about SISs within the context of cybersecurity, in order to allow you to grasp the ideas presented in this book without getting too caught up in the details.

SISs are deemed as the most critical barrier of plant process safety and the last prevention layer against process hazards. Usually, when combined with other engineering and administrative controls, a SIS provides a comprehensive set of safeguards and a layered protection approach as part of a plant's safety philosophy to control risk to **As Low As Reasonably Practicable (ALARP)** or **As Low As Reasonably Achievable (ALARA)**, taking into account social and economic factors. However, these measures are separate from those of a **Basic Process Control System (BPCS)**, which is used for process control. This is the key differentiator between an SIS and a BPCS.

According to the **International Electrotechnical Commission (IEC)** definition, SISs are built to achieve three key objectives:

- To safely and gracefully (or partially) shut down a process when something goes wrong (i.e., a deviation from the norm)
- To let a process run when safe conditions are met
- To respond in a timely manner to prevent **Emergency Shutdown (ESD)**, mitigate **Fire and Gas (F&G)**, or minimize the consequences of a hazard

The term SIS typically consists of multiple elements. It includes, but is not limited to, sensors or detectors to monitor process conditions, logic solvers or controllers to process input signals, and final elements (such as valves or actuators) to perform operations and communication networks that facilitate the exchange of information. These components work together to ensure that the process remains within safe operating limits and to initiate an appropriate response when a safety-critical situation arises.

SIS elements

As depicted in the following illustration, an SIS consists of three key elements:

Sensor   Logic solver   Final element

Figure 1.1 – SIS elements

Let's discuss them further:

- **Sensor**: The sensors (or transmitters) are used to measure the process variable conditions and detect any hazardous conditions in the process.

  Here are some common types of SIS sensors used in process industries:

  - **Pressure transmitters**: Utilized to measure the pressure of gases or liquids in pipes or vessels

  - **Temperature transmitters**: Employed to gauge the temperature of liquids or gases in vessels or pipes

  - **Level transmitters**: Used to measure the level of liquid in tanks or vessels

  - **Flow transmitters**: Widely deployed to measure the velocity of liquids or gases in pipes

  - **Gas detectors**: Employed to ascertain the presence of hazardous gases in the environment, such as carbon monoxide and hydrogen sulfide

  - **Flame detectors**: Used to detect the presence of flames, such as those caused by a fire

  - **Smoke detectors**: Utilized to detect the presence of smoke, which can indicate the presence of a fire

  - **Motion sensors**: Used to detect the movement of equipment or materials in a process, and can help to identify potential hazards or abnormal conditions

- **Logic solver**: The logic solver is essentially the CPU of the SIS that receives input signals, applies safety logic, and generates output signals to control devices such as valves and actuators. It processes data and makes decisions to ensure the safe operation of a process or industrial plant.

- **Final element**: The final element of an SIS is a physical device such as an on/off valve or actuator. It receives output signals from the logic solver and executes the necessary actions to maintain the safety of the plant.

A safety function is part of a system that can have several subsystems and elements:

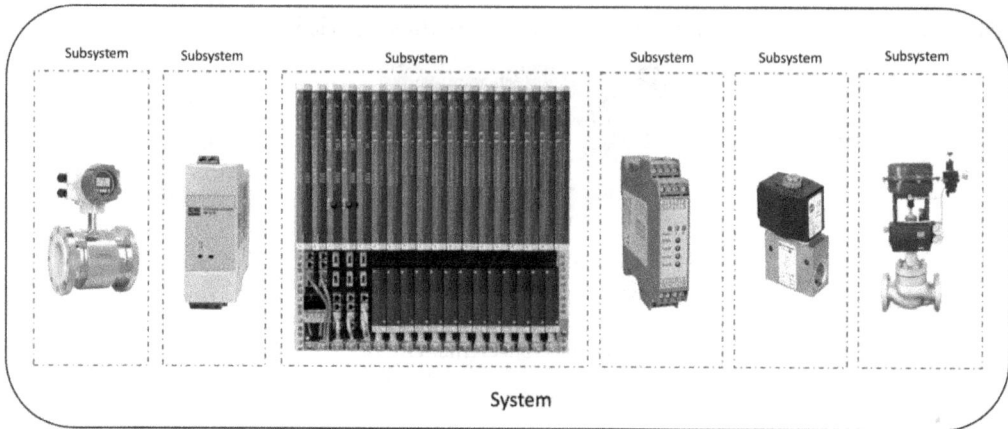

Figure 1.2 – Example of a system and subsystems

Like any complex system, an SIS can experience failures. There are several types of failures that can occur in an SIS, including the following:

- **Random hardware failures**: These are spontaneous failures at random times, which result from one or more possible degradation mechanisms in the hardware – for example, the aging of electronic components, mechanical failure of relays or solenoids, and so on.

- **Software failures**: SISs typically rely on software to perform complex calculations, monitor process data, and control final elements. Software failures can occur due to programming errors, memory leaks, or other issues.

- **Systematic failures**: These are when a pre-existing fault occurs under particular conditions and can only be eliminated by removing that fault by modification of the design, process, procedures, documentation, or other relevant factors.

  Examples of systematic failures could be a hidden fault in the design or implementation of software as well as hardware, an error in the design specifications, user manuals, procedures or security operational procedures (SOPs), and so on. It can occur in any lifecycle phase activity.

- **Configuration errors**: SISs must be carefully configured to ensure that they perform their intended functions correctly. Configuration errors can occur due to human error, deviations or derogations, misinterpretation of specifications, or as a result of changes made to the system that are not properly tested.

- **Environmental factors**: SISs can be impacted by environmental factors such as temperature, humidity, and vibration. For example, extreme temperatures can cause electronic components to malfunction, and vibrations can cause wires or other connections to become loose.

- **Cybersecurity threats**: SISs are increasingly integrated with a BPCS, which increases their attack surface and makes them more susceptible to cyber risks. This can affect both process integrity and system availability.

An SIS can operate in four distinct states that are defined by the state I/O signals originating from the system, as presented in the following table:

SIS state	Process status
OK	Process is available
Safe	Process has tripped
Dangerous	Process is available but not protected
Intermediate	Process is available and SIS is available, but it is time to repair it

Table 1.1 – Different SIS states

> **Important note – deviations and derogations**
>
> **Deviations** typically refer to a departure from the standard performance or prescribed procedures of a system. In functional safety, for instance, a deviation could denote a failure in a safety function or system, resulting in the system not performing as intended. Such deviations could be due to individual component failures, system errors, or security weaknesses. Addressing these deviations necessitates investigating the root cause and devising corrective measures to bring the system back to its standard operating condition. In terms of cybersecurity, deviations could represent any unexpected or irregular activities that could potentially signify a breach or vulnerability threat that requires immediate investigation and remediation.
>
> **Derogations**, on the other hand, represent a formal exemption from a standard or regulation. In the arena of ICS cybersecurity, derogations are often granted when it is impractical to adhere strictly to the standard or when alternative measures provide an equal or higher level of security. Typically, such derogations must be securely controlled, justified properly, and approved by relevant authority figures, ensuring they don't compromise the overall integrity of the system. It's important to note that derogations are not shortcuts or loopholes but are considered flexibilities within the regulatory framework, provided they don't compromise the objective of the standard.
>
> Both deviations and derogations hold immense significance for an ICS's functional safety and cybersecurity. While managing deviations involves identifying, analyzing, and remediating unexpected occurrences, handling derogations involves ensuring any exemptions from standards maintain the requisite level of safety and security.

BPCS versus SIS

SISs are primarily designed to track and sustain the safety of the process and are typically passive and dormant for long periods of time. SISs wait to respond to system demands only when necessary. They use **Safety Instrumented Functions (SIFs)** to execute specific safety-related tasks such as **Emergency Shutdown (ESD)** and **Fire and Gas (F&G)**.

Maintenance and diagnostics are essential in SISs to confirm that the system is functioning properly and reduce the need for manual tests. All SIS modifications after installation require strict compliance with the **Management of Change (MoC)** processes, as even the slightest alteration can have a significant impact.

On the other hand, BPCSs are very dynamic in nature with numerous changes. A BPCS provides oversight over the process with a range of digital and analog inputs and outputs that respond to logic functions, making it easier to detect any malfunctions or failures. However, these systems require frequent changes to ensure accurate process control. BPCSs typically consist of hardware and software components, including sensors, controllers, **Human-Machine Interfaces (HMIs)**, and communication networks. BPCSs often use open standard protocols, such as Modbus and OPC, to communicate with other devices in the plant.

The following figure illustrates the typical components of BCPS and SIS and how they interact from a process perspective:

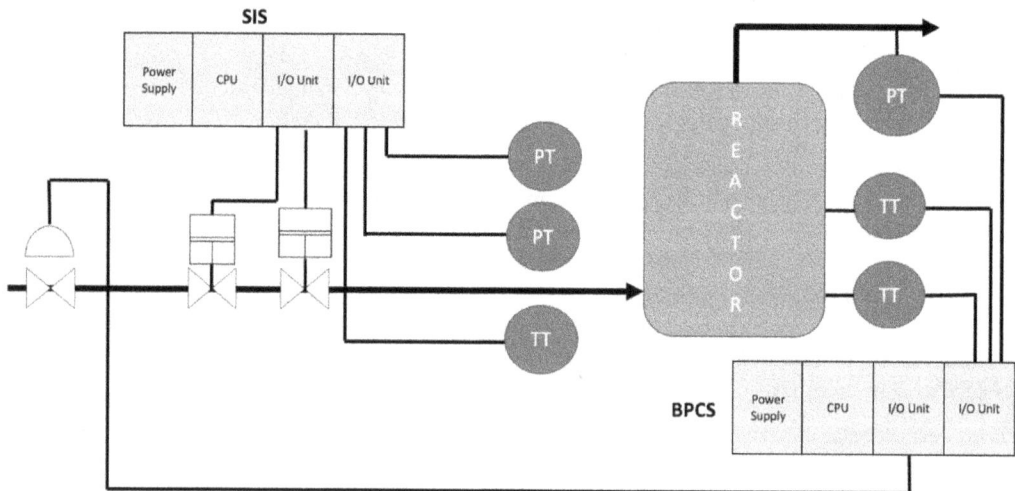

Figure 1.3 – BPCS versus SIS

SIS and BPCS have many similarities, yet their differences lead to different design, maintenance, and integrity requirements.

The implementation of cybersecurity for these systems varies significantly, yet both are susceptible to various threats, including malware, hacking, zero-days, **Man-in-the-Middle (MitM)** attacks, and human errors. Nevertheless, the ramifications of a successful SIS breach can be more severe than in BPCSs, as SISs are responsible for protecting the plant and its personnel from hazardous events. A compromised SIS can lead to the failure of safety functions and potentially catastrophic consequences, such as fires, explosions, and toxic releases. We will explore this further in the next chapter.

SIS applications – where are they used?

SISs are of paramount importance when it comes to protecting process safety. Process plants are beneficial as they can transform raw materials and ingredients into tangible products and goods as part of a complex supply chain. Unfortunately, the techniques used to conduct this conversion can trigger dangerous conditions that, if not efficiently controlled and properly contained, might cause major incidents or top events. Hazardous conditions may be present when dealing with combustible materials such as solids, liquids, gases, vapors, and dust.

In addition, administrative controls and safeguards should be used to address the control of risk.

SISs are deployed for many purposes in petrochemical facilities and pipelines and for other industry-specific needs. Examples of these systems include the following:

- **ESD**: This is a specialized form of control system, created to provide an extra layer of safety for high-risk areas such as oil and gas, nuclear power, and other potentially hazardous environments. Primarily, these systems serve to protect both personnel and the environment if process parameters exceed acceptable levels. By minimizing the potential damage from emergency scenarios such as uncontrolled flooding, the escape of hydrocarbons, and fire outbreaks, ESD systems provide an invaluable service.

The following screenshot presents an example of an ESD system and its components:

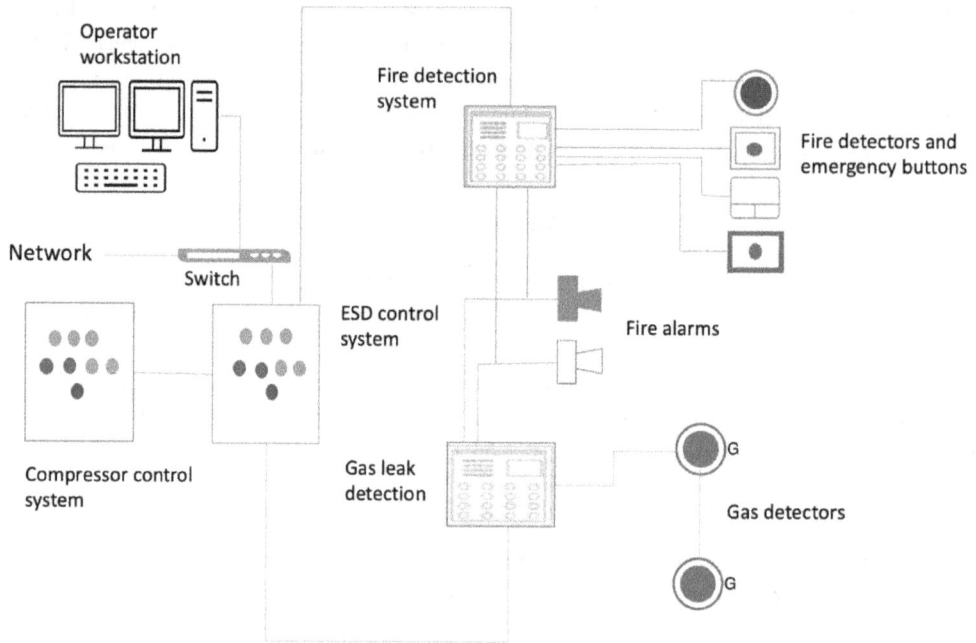

Figure 1.4 – ESD system

The main purpose of ESD can be summarized as follows:

- ESD systems detect unsafe conditions and initiate a shutdown of the process to prevent potentially hazardous situations.

- ESD systems are equipped with sensors that monitor process parameters such as pressure, temperature, level, and flow. If any of these parameters exceed a predetermined limit, the system will initiate a shutdown of the process.

- ESD systems can also be used to activate safety alarms or to stop certain components of the process. This ensures that safety is maintained and potential hazards are avoided.

- **High-Integrity Protection Systems (HIPSs)**: HIPSs are deployed to prevent **Process Shutdown (PSD)** from being affected by any of the destructive factors of overpressure, elevated temperatures, and high-level events. The valves in the HIPS are closed decisively to make the production line secure, and one set of triggers records the observed processes, the logic solver (controller) processes the data, and a few end elements take the safeguarding action by cutting down or stopping the pumps with valves or actuated pumps and circuit breakers that perform the closing (shutdown) operation.

The HIPS serves as the ultimate protection system for the process, and often eliminates the need for pressure release, thereby tending to the environment and mitigating the risks linked to manual handling errors. It also calibrates the overconfidence (high level of trust) that engineers might sometimes have in **Distributed Control Systems (DCSs)** and ESD systems.

Some of the most popular deployments of HIPSs include, but are not restricted to, the following:

- High-integrity pressure protection systems

- High-integrity temperature protection systems

- High-integrity level protection systems

- HIPS interlock systems

The following illustration depicts a typical HIPS deployment for a subsea field environment:

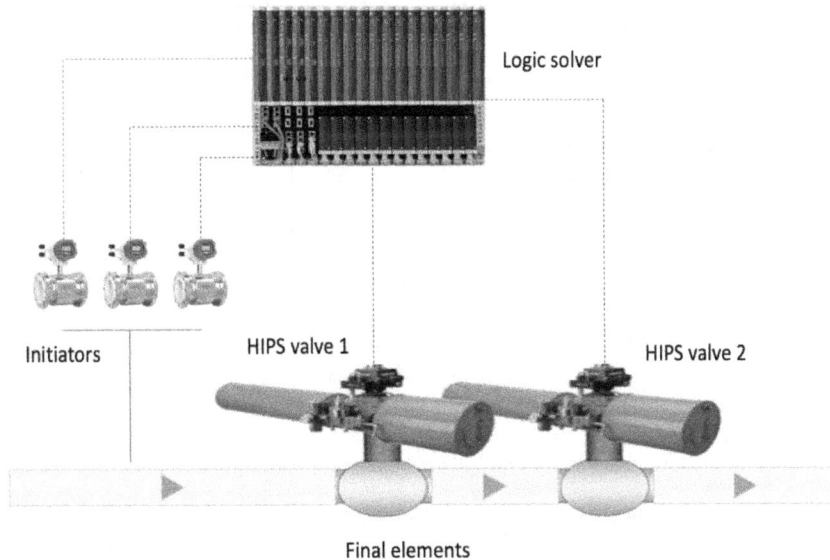

Figure 1.5 – HIPS

- **Burner Management System (BMS)**: This is typically employed to ensure the safe ignition and operation as well as the shutdown of industrial burners when required. This system can be found in many process industries including oil and gas, power generation, manufacturing, and chemical industries, that rely on flame-operated equipment such as furnaces, boilers, and the like. The system is able to keep track of flames with flame detectors, as well as manage igniters, burners, and other actuators such as shut-off valves.

The majority of BMSs are designed with the aim of providing protection against potentially hazardous operating conditions and the admission of fuel that is not suitable. A BMS gives the user important status information and support, while additionally, if there is a hazardous condition, it can initiate a safe operating condition or a shutdown interlock.

According to the **National Fire Protection Association (NFPA) 85 Boiler and Combustion Systems Hazards Code**, a BMS is a control system that is devoted to boiler furnace safety and operator support. This system assures the safe and efficient working of the boiler, thereby contributing to the safety of the facility as a whole.

The chance of fire and hazards will increase significantly without a BMS in place. Organizations nowadays implement BMS in SIS to increase safety and system availability, as well as to remain compliant with sector regulations and the latest industry best practices.

Figure 1.6 illustrates an example of a BMS and its various elements:

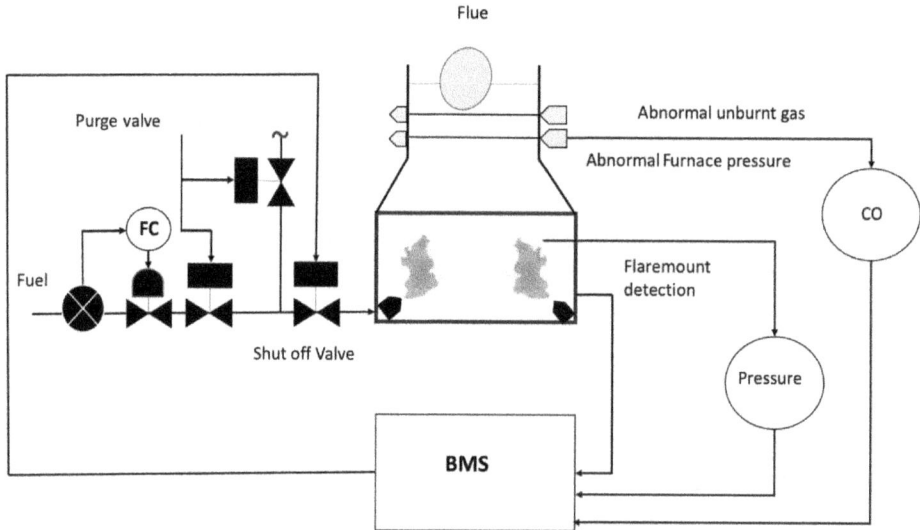

Figure 1.6 – BMS

A list of BMS components, including their functions, can be found in the following table:

Component	Function
Burner	This is where a combination of fuel, oil, and/or gas is mixed with air and ignited to create heat. The process of combustion takes place in multiple burners of large heaters.
FC (flow controller)	This is used to monitor and control the fuel valves and ignitors of the BMS following a sequence that includes processes such as purging, ignition, operation, and shutdown.
Flame detector	This device is used to monitor the absence or presence of a flame and deploy a specific signal to detect it.
Valves	Their primary function is to control and shut off the flow of substances (oil, gas, etc.) into the fuel system.

Table 1.2 – BMS components and functions

It is no surprise that SISs play an essential role within process industries in guaranteeing the safety and dependability of critical operations. A few examples of where an SIS is required to aid in the safeguarding of people, equipment, and the wider environment include the following:

- **Process safety in the chemical industry**: The use of SIS in the chemical industry can be focused on **Health, Safety, and Environmental (HSE)** considerations, and mitigating the consequences of a major accident. For example, an SIS can be used to automatically shut down a process if a critical parameter exceeds a predetermined limit, thereby preventing a catastrophic incident.

- **Power generation**: An SIS can be used in power generation plants to protect critical equipment and processes, such as turbines, boilers, and generators. For example, an SIS can be used to automatically shut down a turbine or generator in the event of an abnormal condition, such as low oil pressure or high temperature, to prevent damage to the equipment and ensure safe operation.

- **Transportation safety**: An SIS can be used in transportation systems, such as railways and pipelines, to detect and mitigate hazardous conditions. For example, an SIS can be used to automatically apply the brakes on a train if it exceeds a certain speed limit or if it encounters an obstacle on the track, thereby preventing a potential collision.

- **Offshore oil and gas production**: An SIS can be implemented in oil and gas environments – including oil fields and offshore platforms – to protect personnel as well as assets from the hazards of explosive gases, fire, and other risks associated with the production process. For example, an SIS can be used to automatically shut down production in the event of a leakage of gas or fire to prevent an explosion or other catastrophic event.

In the next section, we will examine ICS cybersecurity as a new discipline in detail. We will also explore how the IT and engineering communities perceive ICS cybersecurity in their respective fields.

What is ICS cybersecurity?

The term ICS is used in a broad sense to refer to programmable-based devices that are used to control, monitor, supervise, automate, or interact with assets used in continuous, discrete, and hybrid processes in manufacturing, infrastructure, and other commercial and industrial sectors.

At its heart, ICS cybersecurity is about both protecting industrial assets and recovering from system upsets that occur from electronic communications between systems, or between systems and people.

An ICS includes various components, such as the following:

- **Distributed Control Systems (DCS)**
- **SIS**
- **HMIs**
- **Historians**

- **Supervisory Control And Data Acquisition (SCADA)**
- **Programmable Logic Controllers (PLCs)**
- **Remote Terminal Units (RTUs)**
- **Intelligent Electronic Devices (IEDs)**
- **Power Monitoring Systems (PMSs)**
- **Protection relays**
- **F&G**
- **ESD**
- **PSD**
- **BMS**
- **Building Control Management Systems (BCMSs)**
- **Electrical Network Monitoring Control Systems (ENMCSs)**
- **Alarm management systems**
- **Intelligent Asset Management Systems (IAMSs)**
- **Sensors and transmitters**
- **Valves**
- **Drives, converters, and so on**

Establishing a secure baseline for an ICS can be a complex and wide-reaching process as this can cover software, hardware, and communications interfaces. These hardening parameters need to be defined, at the very minimum level, by the following:

- OS security
- Endpoint security
- Embedded device security
- Application software security
- Network security
- Access control (physical and logical)
- Anti-malware
- Security monitoring

Despite certain common attributes, ICS differs from the traditional IT systems that are widely deployed in office and enterprise networks. Historically, ICS implementations were heavily reliant on physical security and lacked interconnection with IT networks and the internet. However, as the trend toward ICS intertwining with IT networks intensifies, this creates a greater need to secure these systems from remote, external threats as well as against adversary and non-adversary threats such as disgruntled employees, malicious intruders, and malicious or accidental actions taken by insiders.

In relation to the CIA's information security model, availability and integrity are given precedence over confidentiality for ICS. The ICS security model is therefore often referred to as an AIC model. In the meantime, **reliability and safety remain top priority**!

The following figure compares the priorities of the ICS security model with the IT information security model:

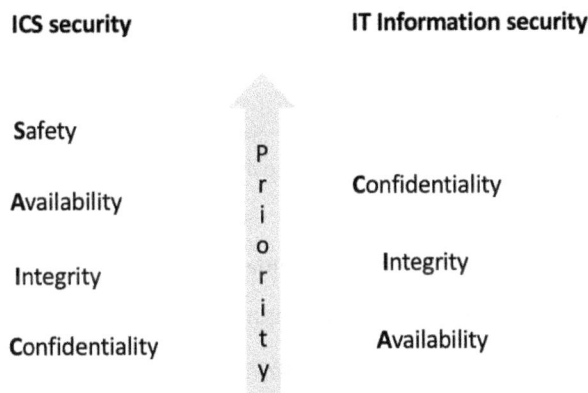

ICS security	Priority	IT Information security
Safety		
Availability		Confidentiality
Integrity		Integrity
Confidentiality		Availability

Figure 1.7 – An ICS versus an IT model

Let's have a closer look at the definition of each element of the (S)AIC triad:

- **Safety**: The assurance from unacceptable risk.
- **Availability**: The ability of a system or asset to be accessed and used by an authorized user when required.
- **Integrity**: The assurance that a system or asset is accurate and complete. It also refers to the assurance that the system or asset can only be modified by an authorized user.
- **Confidentiality**: The assurance that a system or asset is only accessible to an authorized user and is kept secure from unauthorized access. It also refers to the assurance that information within the system or asset is only accessible to an authorized user.

The increasing convergence of business and plant floor systems, emerging standards such as the **International Society of Automation**'s ISA/IEC-62443 and the **National Institute of Standards and Technology**'s NIST 800-82 series, and emerging regulatory requirements in a number of countries, all point toward a growing awareness of the susceptibility of the modern industrial process to cybersecurity threats.

Considering the potentially dangerous safety consequences that can occur as a result of these failures, today's plants need to clearly understand the actual risks – and how best to mitigate these risks – in order to maintain safe and reliable operations.

The potential implications of ICS security breaches encompass a wide range of damaging consequences that might include, but are not limited to, asset, financial, environmental, and reputational damage:

- Compromise and unauthorized disclosure of confidential data to the public
- Tampering of system reliability or integrity of process data and production information
- **Loss of View (LoV)** and **Loss of Control (LoC)**
- Process abuse and corruption that could bring about degraded process efficiency, poor product quality, diminished manufacturing capability, impaired process safety, or environmental release
- Damage to assets
- Health implications including injuries and fatalities
- Demeaned and negative reputation and public trust
- Breach of contractual and regulatory obligations (such as clients, partners, and regulators)
- Impact on national security and critical infrastructures

The following consequences have already occurred within ICS installations including SIS:

- Manipulation of process data or setpoints
- Unauthorized changes to commands or alarm thresholds
- Erroneous information being passed on to operators (loss or manipulation of view)
- Software or settings being tampered with and interference with safety systems, all of which could have far-reaching and potentially fatal consequences

How do IT and engineering communities perceive ICS cybersecurity?

The IT and engineering communities are increasingly aware of the need for ICS cybersecurity. As ICS become ever more connected and automated, they also open themselves up to greater risk of cyberattacks. To address this, both communities are now developing a range of solutions and working closely to protect these systems from emerging threats.

While both communities view ICS cybersecurity from different angles and perspectives – due in large part to the historical gap that exists between IT and ICS as well as differing priorities – they have come to recognize the need to bridge the gap in order to tackle the increasing challenges facing industrial facilities. As a result, a new discipline has emerged that combines the best of both engineering and cybersecurity practices.

For example, engineers are typically more focused on the physical process of an ICS, such as the hardware and software, while IT professionals are more concerned with the network and data security aspects.

A more comprehensive approach to ICS cybersecurity can be achieved by combining both engineering and IT practices. This includes both the physical and the digital components of the system to ensure that the assets are secure from cyber threats.

The following sections will dive into the distinct aspects of international standards for cybersecurity and safety.

Exploring relevant cybersecurity and functional safety standards

Industry associations and governments have established various cybersecurity and functional safety standards in recent years, providing mandatory guidance and regulations for compliance. Furthermore, these standards are regarded as industry best practices for ensuring the safety and reliability of numerous process industries.

These standards are issued by the IEC, thus many countries have superseded their own national requirements and implemented these standards instead. This has provided substantial operational leverage for businesses with operations in multiple countries, as the global standards allow for a single standard to be applied throughout the organization.

This section will provide an overview of SIS-applicable standards with a brief description of the relevant security controls. For other functional safety requirements not outlined here, we recommend that you review the applicable IEC standards.

The IEC provides two renowned, widely used functional safety standards – IEC 61508 and IEC 61511:

- IEC 61508 is a general safety document from the IEC that provides an overarching framework for achieving functional safety in safety-related systems for many industries and applications. IEC 61508 is used as a foundation for sector-specific functional safety standards including IEC 61511, IEC 61513, ISO 26262, and IEC 62304.
- IEC 61511 is a dedicated standard that is primarily focused on process industries and is based on IEC 61508.
- IEC 62304 covers software safety classification, while ISO 26262 is about road vehicles' functional safety.

The following diagram depicts the most widely used industry functional safety standards:

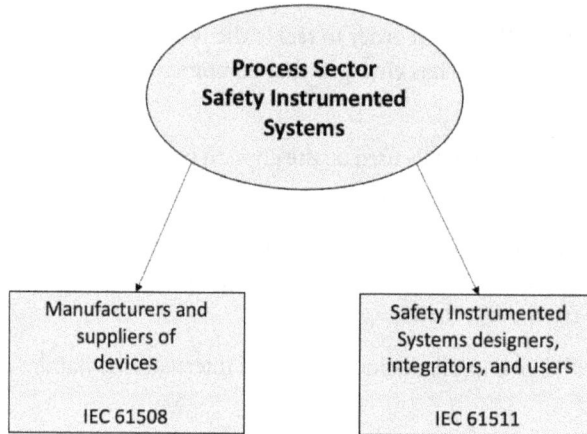

Figure 1.8 – Scope of IEC 61508 and IEC 61511

The scope of IEC 61508 and IEC 61511 can be described as follows:

- **IEC 61511 – Functional safety – Safety Instrumented Systems for the Process Industry Sector**

    IEC 61511 is a global norm prescribing requirements and guidance for the formation, execution, and operation of SIS for the process industries with a spotlight on the end users. The standard encompasses the overall safety lifecycle of an SIS, including cybersecurity requirements as a part of functional safety and risk management as stipulated by IEC 61508.

    In terms of safety and cybersecurity intersection, the IEC 61511 standard (edition 2) was amended in 2016, with clause 8.2.4 outlining the need for conducting a cybersecurity risk assessment to determine the presence of any potential security weaknesses or vulnerabilities on the SIS. To this end, users of the IEC 61511 standard are directed to seek guidance related to SIS security from the IEC 62443 standards and ISA TR84.00.09.

    IEC61511-1: 2016 edition 2 `https://webstore.iec.ch/publication/24241` clause 8.2.4 mandates a thorough examination of security risks to pinpoint any vulnerabilities within the SIS. This assessment should encompass the following:

    - Defining the devices under scrutiny (including the SIS, BPCS, or any connected devices)

    - Identifying potential threats capable of exploiting vulnerabilities, leading to security breaches (ranging from deliberate attacks on hardware and software to inadvertent errors)

    - Assessing the potential repercussions of security breaches and estimating their likelihood

    - Addressing various project phases, including design, implementation, commissioning, operation, and maintenance

- Determining any additional measures required to mitigate risks

- Outlining the steps taken to mitigate or eliminate identified threats, or providing references to relevant information

- **IEC 61508 – Functional Safety of Electrical/Electronic/Programmable Electronic Safety-Related Systems**

  IEC 61508 is a standard series of functional safety, which applies throughout the lifespan of **Electrical, Electronic, and Programmable Electronic (E/E/PE)** systems and products. This set of regulations encompasses parts of devices and equipment that perform automated safety characteristics; these components may include sensors, control logic, actuators, and microprocessors.

  The uniform technical approach mandated by IEC 61508 can be applied to all safety systems within the electronics and related software industries, regardless of sector. Not only does this horizontal standard target suppliers of safety systems but it can also be used to some extent by those that provide equipment for these safety systems. Furthermore, IEC 61508 sets out four different **Safety Integrity Levels (SILs)** to determine the success of a system in meeting its specified safety objectives. These SILs are dependent on the robust analysis of the potential risks and hazards of a device, as well as on the consequent likelihood and severity of any such hazard.

  Clause 7 of the standard, specifically titled *Realization of the Safety Function*, includes criteria for implementing safety functions in an SIS, as well as requirements and guidance for addressing cybersecurity risks. Therefore, although the IEC 61508 standard does not concentrate exclusively on cybersecurity, it does provide recommendations for mitigating cybersecurity risks in an SIS.

As for ICS cybersecurity, common ICS security-related standards include the following:

- **ISA/IEC 62443 – Security of Industrial Automation and Control Systems**

  The IEC 62443 series provides a structural foundation that encompasses the safety of **Industrial Automation and Control Systems (IACSs)** including SIS. This set of standards currently consists of 13 documents that address topics such as developing a proper IACS security program and system design requirements for securely integrating control systems. Additionally, ISA TR84.00.09 builds on the work of ISA99 for IEC 62443 and examines defensive measures to reduce the chance of a breach that may compromise the SIS's performance. This technical report also furnishes criteria for warding off external and internal security threats and outlines ways to meet the requirements of IEC 61511.

The following diagram provides an overview of the IEC 62443 standards series and key areas of focus:

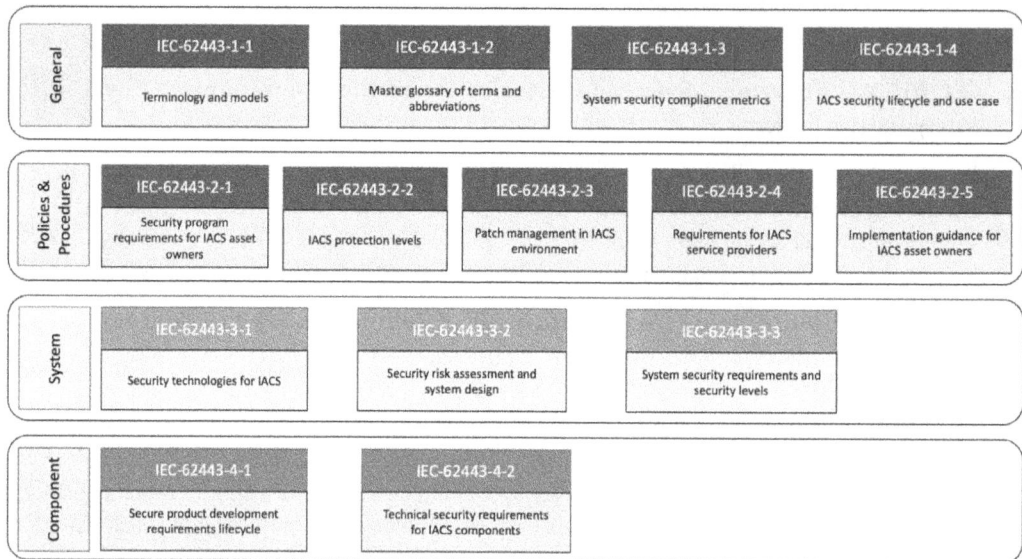

Figure 1.9 – Structure of the IEC 62443 series

- **NIST 800-82 – Guide to Industrial Control Systems (ICSs) Security**

  SP 800-82 from the National Institute of Standards and Technology affords insight into enhancing the security of ICSs. This includes SCADA, DCS, and PLCs, while also handling their varied specifications as well as safety prerequisites. It offers an excursus on ICSs and their general system layouts, pinpoints potential threats, and prescribes countermeasures to cut down the related risks, including SIS. NIST 800-82 emphasizes the importance of risk management in ICS security by providing guidance on conducting risk assessments, identifying threats and vulnerabilities, and developing risk mitigation strategies.

- **NRC regulation 5.71 – Cyber Security Programs for Nuclear Power Reactors**

  The US Nuclear Regulatory Commission's 10 CFR 5.71 regulation stresses the significance of cyber defense in the architecture and running of systems that are safety-critical. It requires licensees to build and execute digital safety-related systems with the highest levels of assurance, making sure that they are resilient to cyber intrusions that could jeopardize their safety functions. This regulation also mandates licensees to put cybersecurity programs into place that incorporate certain measures to both manage cyber threats and maintain system dependability and consistency in the long run.

Revision 1 of NRC regulation 5.71 provides required guidance on **Defense-in-Depth (DiD)** practices based on international standards such as the NIST 800 series and **International Atomic Energy Agency (IAEA)** cybersecurity guidance. This version provides insight into concerns raised from cybersecurity reviews, trends in the industry, emerging legislations, and disruptive technologies as well as outreach programs including lessons learned from cybersecurity incidents.

- **NEI 08-09 – Cyber Security Plan for Nuclear Power Reactors**

  NEI 08-09 is a high-level security plan (or strategy) with a layered architecture and a variety of security controls based on the NIST SP 800-82 and NIST SP 800-53 standards. This strategy ensures that systems and networks linked with safety-related operations are protected against cyberattacks that could potentially harm their mission critical functions.

- **NERC CIP**

  The North American Electric Reliability Council Critical Infrastructure Protection (NERC CIP) is a regulation to monitor, enforce, and manage the cybersecurity of the Bulk Electric System (BES) in North America. This set of standards is intended to identify and protect vulnerable assets that can influence the reliable supply of electricity throughout the continent's BES. The CIP framework is designed to ensure the security of the CI.

  Requirements CIP-002-5.1a and CIP-005-6 under NERC CIP focus on the identification and protection of cybersecurity management for safety systems. These standards mandate that responsible organizations must recognize and record details of safety systems. These are described as systems and equipment that are essential for detecting, preventing, or mitigating scenarios that might cause significant disruptions or hinder the safe shutdown of the bulk electric system.

  Here is a high-level overview of these standards:

  - **CIP-002-5.1a BES Cyber System Categorization**: This categorizes BES cyber systems and their associated assets to tailor cybersecurity measures appropriately, based on the potential impact that damage, unauthorized access, or misuse could have on the BES's reliability.

  - **CIP-005-6 Electronic Security Perimeter**: This defines a controlled boundary around networks where critical cyber assets are connected, controlling access to these networks. The goal is to regulate electronic access to BES cyber systems and establish a secure perimeter to prevent actions that could disrupt or destabilize the BES.

In the next section, we will discuss the various stages of the functional safety lifecycle as well as the high-level cybersecurity phases that are crucial to safety critical systems. We will also explore the common processes and methods that are used in each phase as well as their importance in ensuring safe operations.

Examining the safety and cybersecurity lifecycle

This section will cover the safety and cybersecurity lifecycle, exploring different functional safety phases as well as common practices to reduce risk.

Safety lifecycle

Recent safety standards pertaining to SIS have a core concept of the **Safety Lifecycle (SLC)**. This engineering process is built to ensure a comprehensive level of safety from analysis to implementation, covering the operational and maintenance phases of the system. By adhering to the SLC's rules and regulations, industrial automation systems are ensured to be able to efficiently reduce the industrial process risk.

In addition, the SLC offers the following baselines:

- A routine, steady architecture for the definition, planning, establishment, and upkeep of an SIS

- A solid foundation for **Risk Assessment Methodology (RAM)** techniques

- An SIS management system, and the **Key Performance Indicators (KPIs)** expected of each safety instrumented function

The following diagram illustrates the required steps and phases that can be found as part of the SIS SLC:

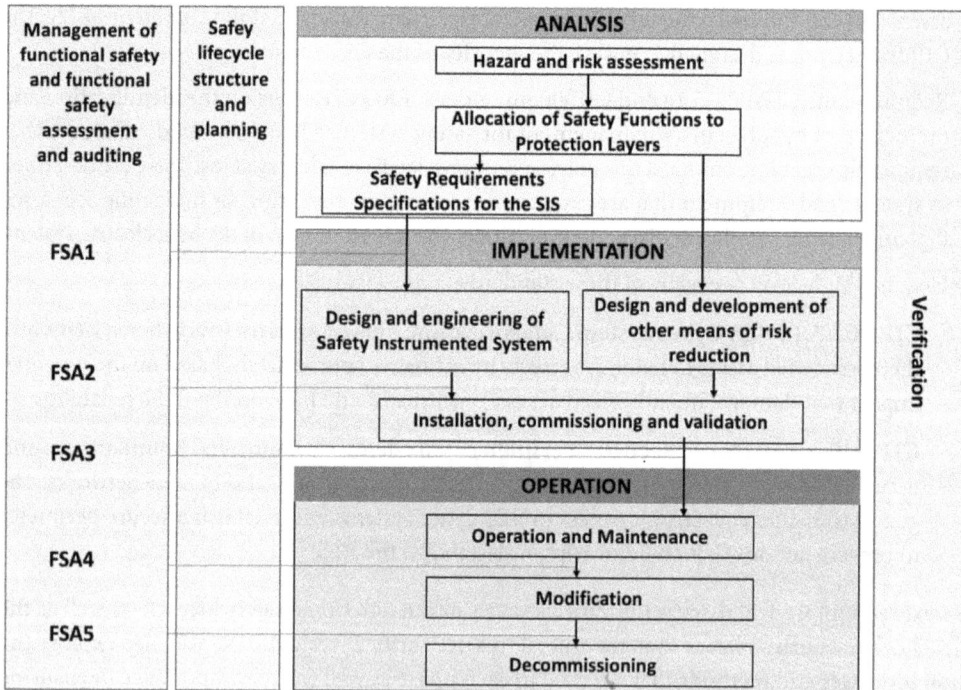

Figure 1.10 – ISA-84.00.01-2004 SIS SLC

The preceding diagram provides a high-level overview of the SLC's main phases that we will cover briefly here:

- **Analysis phase**

  This phase systematically identifies hazards, assesses risks, and defines the safety requirements of the system in order to design and implement effective security instruments (SIFs) that can minimize the risks associated with the system.

  Furthermore, this phase also involves developing a safety concept and safety functions to determine the necessary safety integrity measures and reduce the risk of hazards to an acceptable level. In certain cases, hazards will be found to be within an acceptable range, and as such, no further mitigation is required.

  Therefore, no SIF is warranted. However, in other instances, a risk mitigation measure is needed, and its effectiveness is determined by its **Safety Integrity Level (SIL)**.

- **Implementation phase**

  Once the SIFs have been identified and documented, work can commence on the design. This includes the selection of suitable vendors for the sensor, logic solver, and final element, as well as the determination of whether to include redundancy for high safety integrity, to minimize false trips, or both. Subsequently, after the selection of products and their associated components, the design should review the safety philosophy and any known constraints as identified and provided in the **Safety Requirements Specification (SRS)**. As the SIS is designed to not be activated, it is essential that it be inspected and evaluated thoroughly at predetermined intervals.

- **Operation phase**

  The operation phase is the final phase of the SIS functional SLC. During this phase, the SIS is fully operational and is used for its intended purpose. This phase includes activities such as the ongoing monitoring, maintenance, and verification of system effectiveness. The goals of this phase are to ensure that the SIS continues to perform its intended function and to identify and address any potential issues that could negatively impact safety. This phase is critical for maintaining the safety of the system and ensuring its continued reliability.

  If there are any modifications to be carried out, these must strictly follow the MoC protocol of the organization and a **Stage 5 Functional Safety Assessment (FSA5)** should be conducted. Regular audits must also be part of this essential lifecycle phase.

  As part of this phase, questions related to design and maintenance, the management of change processes, and so on must be addressed within the **Pre-Startup Safety Review (PSSR)**. Examples of these questions include but are not limited to the following:

  - Does the system comply with all the specifications outlined in the SRS?

  - Have the SIL targets and **Mean Time to Failure (MTTF)** targets been achieved for all SIFs?

  - Are all the requirements of the SIS SLC being effectively completed?

- Is all equipment configured in accordance with the manufacturer's safety manual?

- Has a **Hazard and Risk Analysis (H&RA)** been carried out and have any recommendations been implemented?

- Have the recommendations from any **Functional Safety Assessment (FSA)** been resolved?

- Has a cybersecurity evaluation been conducted?

- Is there an established schedule for periodic inspections and tests for each SIF?

- Have the maintenance procedures been established and validated?

- Is there an established procedure for managing changes?

- Is there a security patch management strategy enforced?

- Are the operation and maintenance teams trained, certified, and qualified for the work?

Only once the aforementioned questions have been addressed adequately can we move on to the startup and operation can continue.

As depicted in *Figure 1.9*, the FSAs, as part of the management of functional safety and functional safety assessment and auditing, are conducted throughout the lifecycle phases:

- **FSA1**: The aim of performing a **Functional Safety Assessment (FSA1)**, once the analysis step has been concluded and the SRS has been created, is to detect any possible safety risks.

- **FSA2**: Once the SIS's detailed design and engineering have been finished, it is necessary to carry out a **Functional Safety Assessment (FSA2)**.

- **FSA3**: Prior to SIS startup and after installation, commissioning, and **Site Acceptance Test (SAT)**, a systematic – and mandatory – SIL validation shall be conducted to fulfill functional safety standard requirements.

- **FSA4**: System operation and maintenance must be conducted by personnel who are qualified and have demonstrable experience from past projects.

    This is an essential requirement. Regular **Stage 4 Functional Safety Assessments (FSA4s)** must be performed to verify the following:

    - The alignment of ongoing activities with the initial design assumptions

    - Full compliance with the safety management and verification requirements stipulated in IEC 61511

- **FSA5**: FSA5 shall be carried out before the modifications. Once the modification activity is complete, another FSA5 shall be required to assess and confirm that the necessary modification is meeting the safety integrity requirements.

Cybersecurity lifecycle

The cybersecurity lifecycle shares strong similarities with the SLC in terms of risk reduction, yet they differ from one another due to their separate design by different communities, each with its own terminologies, contexts, and ways of working.

IT security professionals prioritize dealing with immediate threats, whereas process safety engineers are chiefly concerned with much longer lifespans of up to 10 years. Historically, the role of IT within industrial networks has been focused primarily on data (historian replication in IT) access, support for communication interfaces, or access to tools.

Many forward-focused organizations are now attempting to change the culture and bring these two communities closer together through the formation of new operating models, with the aim of enhancing collaboration and jointly confronting the increasing cyber risk that threatens organizations globally.

With the industry approaching the cybersecurity lifecycle in so many different ways, we will focus solely on industrial standards such as IEC 62443 and NIST, as these include ICS practical guidance. We will explore these further in later sections of this book.

> **Important note**
>
> It is important to emphasize that compliance and security are not the same. The proposed standards provide guidance and advice regarding certain security controls that have been adopted for general use in ICS environments. Nevertheless, no standard is capable of accounting for all the specifications of your company's business processes. Therefore, it is essential to be cautious when implementing these standards for ICS security projects and to remember that adhering to standards does not guarantee your security.

Summary

In this chapter, we have discussed the fundamental concepts that shape functional safety and ICS cybersecurity with a great emphasis on risk mitigation. We have covered the components, standards, terms, and practices that are currently used in multiple industries. Furthermore, we have examined the functional safety and cybersecurity life cycles, including their similarities and discrepancies.

We now have a strong foundation to dive more deeply into practical SIS cybersecurity. The next chapter will explore the need to protect safety processes against emerging cyber threats and will provide examples of recent security incidents that have primarily targeted SIS.

Further reading

- *NIST SP 800-82 Rev 2, Guide to Industrial Control Systems (ICS) Security*:

  `https://nvlpubs.nist.gov/nistpubs/SpecialPublications/NIST.SP.800-82r2.pdf`

- *U.S. Nuclear Regulatory Commission, Regulatory Guide*:

  `https://www.nrc.gov/docs/ML0903/ML090340159.pdf`

- *Cyber Security Programs for Nuclear Power Reactors*:

  `https://www.federalregister.gov/documents/2023/02/13/2023-02941/cyber-security-programs-for-nuclear-power-reactors#:~:text=RG%205.71%2C%20Revision%201%20is,computer%20and%20communication%20systems%20and`

- *North American Electric Reliability Corporation, Cyber Security Standards CIP-002-1 through CIP 009-1*:

  `https://www.nerc.com/pa/Stand/Pages/Cyber-Security-Permanent.aspx`

- *ANSI/ISA-61511-1-2018, Functional Safety – Safety Instrumented Systems for the Process Industry Sector – Part 1: Framework, definitions, system, hardware and application programming requirements*:

  `https://www.isa.org/products/ansi-isa-61511-1-2018-iec-61511-1-2016-amd1-2017-c`

- *ISA-TR84.00.09-2017, Cybersecurity Related to the Functional Safety Lifecycle*:

  `https://www.isa.org/products/isa-tr84-00-09-2017-cybersecurity-related-to-the-f`

- *ISA-62443-2-1-2009, Security for Industrial Automation and Control Systems Part 2-1: Establishing an Industrial Automation and Control Systems Security Program*:

  `https://www.isa.org/products/isa-62443-2-1-2009-security-for-industrial-automat`

- *Final Elements in Safety Instrumented Systems, IEC 61511 Compliant Systems and IEC 61508 Compliant Products*: `https://www.exida.com/Books/final-elements-in-safety-instrumented-systems`

- *ISA - Safety Instrumented System Design: Techniques and Design Verification*: `https://www.isa.org/products/safety-instrumented-system-design-techniques-a-1`

- *ANSI/ISA-84.00.01-2004 Part 1 (IEC 61511-1 Mod) - Functional Safety: Safety Instrumented Systems for the Process Industry Sector - Part 1: Framework, Definitions, System, Hardware and Software Requirements*:

  `https://webstore.ansi.org/preview-pages/ISA/preview_ANSI+ISA+84.00.01-2004+Part+1.pdf`

- *ISA-62443-1-1-2007 Security for Industrial Automation and Control Systems Part 1-1:Terminology, Concepts, and Models*:

  `https://www.isa.org/products/isa-62443-1-1-2007-security-for-industrial-automat`

- *IEC 61508; Functional safety of electrical/electronic/programmable electronic safety-related systems, IEC, 1998, 2000*:

  `https://webstore.iec.ch/publication/5515`

- *IEC 61511-1:2016 Functional safety - Safety instrumented systems for the process industry sector - Part 1: Framework, definitions, system, hardware and application programming requirements*: `https://webstore.iec.ch/publication/24241`

- *ISO 26262-1:2018, Road vehicles – Functional safety – Part 1 Vocabulary*:

  `https://www.iso.org/obp/ui/en/#iso:std:iso:26262:-1:ed-2:v1:en`

- *IEC 62304 (IEC 62304:2006), Medical device software — Software lifecycle processes*:

  `https://www.iso.org/obp/ui/en/#iso:std:iec:62304:ed-1:v1:en`

2
SIS Evolution and Trends

The use of **Safety-Instrumented Systems** (**SISs**) has long played an important role in the critical processes of industrial facilities and military operations, and for decades, their use has extended to other sectors. This widespread adoption has seen SIS technology evolve in tandem with rapid technological advancements, from analog and serial systems right up to the currently implemented fully digital Ethernet-based systems that are now commonplace and widely implemented.

However, as technologies have progressed and evolved, new cyber risks and challenges have also developed, placing many of the technologies that we are now so deeply reliant upon under great threat.

This chapter will comprehensively explore the history, evolution, and current state of SISs in order to help you gain a better understanding of the need for protection against cyber risks. We will examine various **Industrial Control System** (**ICS**) cyber risk trends and their evolution, as well as the intersections that exist between safety and cybersecurity. We will reveal some of the main SIS technologies found in the majority of plants, and we will explore the ICS threat landscape, looking at relevant cybersecurity incidents, case studies, and the associated lessons learned. Lastly, this chapter will finish with a review of present-day ICS cybersecurity solutions.

Over the pages to come, you will gain a much broader grasp and awareness of SISs, the threats posed, and the practical tactics and tools now available to appropriately protect them. This knowledge will be useful not only to regular users, such as the engineers and security practitioners of SISs, but also to organizations and industries that need to ensure their safety systems are robust and up to date. Additionally, learning from the cybersecurity incidents explored within this chapter will help you to develop a greater understanding of risk management, cyber hygiene, and the emerging ICS threat landscape.

This chapter will focus on the following topics:

- The history and evolution of SISs

- The need for protecting SISs

- ICS cyber risk trends and evolution

- The intersection of safety and cybersecurity

- ICS threat landscape

- ICS cybersecurity incidents and lessons learned

The history and evolution of SISs

The history of SISs has been largely shaped by the impact of a number of major safety disasters, including Piper Alpha, Milford Haven, and Deepwater Horizon, among others. These high-profile, damaging incidents have played a significant role in shaping the future of the industry and the implementation of regulations and standards within the process industries. However, when it comes to cybersecurity, a focus on SISs still remains relatively new for ICSs. Nevertheless, it is continually evolving.

According to statistics, the evolving cyber risk trends indicate that ICS environments, including SISs, are becoming an increasingly attractive target for various threat actors. This can be attributed to factors such as the ease of access to information, challenges within the supply chain, and a shortage of qualified personnel who are capable of effectively managing ICS security and promptly mitigating any potential threats.

Moreover, the digital and physical realms become ever more intertwined with each day that passes, and the aftermath of malware and zero days is now having a highly damaging physical impact. Sadly, the huge advantages that digitalization brings in terms of convenience and organizational effectiveness is somewhat countered by the threats to peoples' safety and protection.

There have been multiple concrete examples of cybersecurity incidents and near misses in various industrial sectors in recent years: Triton, Colonial Pipeline, and BlackEnergy are just a few examples of the better-known incidents. As organizations strive to keep pace with the ever-evolving threat landscape, they have had to disclose these incidents publicly; however, it's worth noting that while these incidents have all raised serious concerns, only the Triton incident has been confirmed as a direct attack of targeting an SIS.

On the other hand, attribution remains a big challenge in the world of cyberspace; it is extremely difficult to determine the source of most of these campaigns or targeted attacks. There are simply so many potential factors involved, and these can be carried out by various threat actors, including cybercriminals, nation states, hacktivists, insiders, or even unskilled "script kiddies."

While the ICS security community is undoubtedly learning about the necessary tactics and artifacts that already exist, these past incidents (or top events) and near misses continue to highlight the critical importance of implementing a comprehensive ICS cybersecurity strategy. Given the rapid transformation taking place within ICS environments, it is more vital than ever to enhance the reliability and resilience of SISs against malicious cyberattacks.

The concept of SISs originated in the process industries in the early 1970s in response to a series of process safety incidents that unfortunately resulted in both loss of life and property damage. This led to the development of industry standards and guidance documents such as ANSI/ISA-S84 and IEC 61511. Since then, major vendors (globally) have continually evolved and improved the technology behind their individual SIS products, from pneumatic and electrical systems in the early stages to the much more digitally intelligent devices and systems that are widely used today.

The first generation of controllers relied upon mechanical hardware and pneumatic or hydraulic devices. These systems, which were slow to respond, had limited capabilities and were complex to diagnose and maintain. It required costly replacement and repairs, making them no longer fit for purpose. The digital SISs of the last 40 years have, instead, relied upon electronic sensors, communications networks, and digital controllers – all of which have allowed for much more rapid response times, easier maintenance and diagnostics, and enhanced capabilities. Digital SISs can also be seamlessly integrated with essential control systems, allowing for much more efficient process control.

The following diagram depicts the evolution of SIS controllers to date:

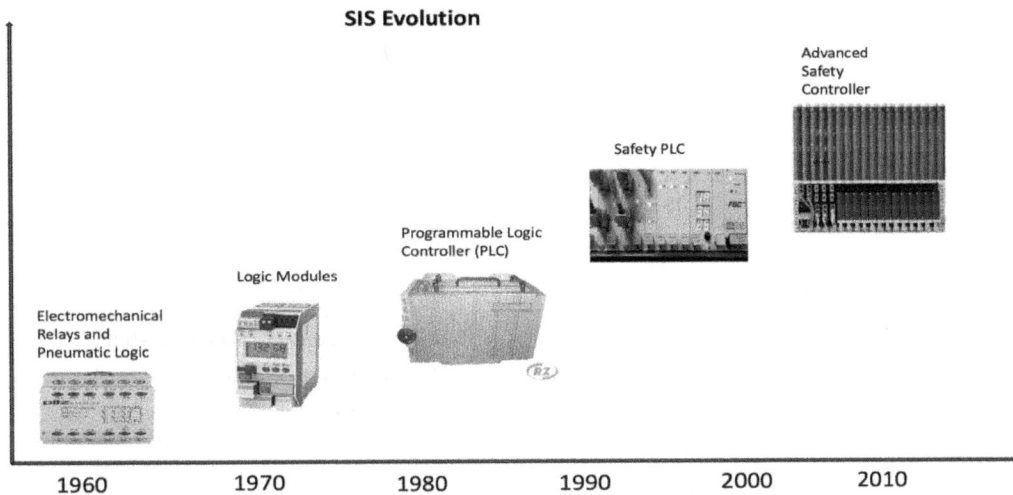

Figure 2.1 – Evolution of SIS controller

The earliest SISs were developed in the late 1970s and early 1980s. These early systems were based on serial communication protocols and had limited computing power, but they could provide basic safety-related diagnostics and alarms. As computers evolved, so did the capability of SISs, and by the mid-1980s, **intelligent** SISs based on Ethernet became widely available. In terms of market share, serial-based SISs represented most of the market throughout the 1980s and early 1990s. However, by the late 1990s, the emergence of Ethernet-based systems had started to make a real impact, with Ethernet-based SISs reaching a 25% market share by the early 2000s.

Initially, the primary industry using SISs was process control, which needed efficient and reliable control systems. By the mid-1990s, however, safety-critical systems were being implemented in a much wider range of industries, including automotive, aerospace, railway, nuclear, and power generation. These systems typically used Ethernet-based SISs, as they provided more reliable communication, greater computing capability, and the ability to integrate with other types of control systems. Today, Ethernet-based SISs have all but replaced serial-based systems, with Ethernet-based SISs now accounting for more than 90% of the market. This increase in the use of Ethernet-based SISs has been driven by the need for enhanced safety, greater heightened convenience, and reliability. As a result, SISs are now used in virtually every industry where safety is considered to be a critical factor.

The following table provides some of the key characteristics of using analog or digital:

	Benefits	Limitations
Analog	• Easier to install and modify • Does not require specialized training to operate • Relatively lower upfront costs	• Limited capabilities due to mechanical constraints • Slow response time • Complex maintenance and diagnostics • High cost of repairs and replacements
Digital	• Faster response time • Easier maintenance and diagnostics • Enhanced capability • Easier integration with other control systems	• More expensive upfront cost • Can potentially require specialized knowledge for setup or repairs • Additional precautions and systems needed for guarding shift-out timeline accuracy

Table 2.1 – Analogue vs. digital characteristics

SIS technology is a key factor when it comes to the protection and safety of people, plants, and equipment. Major vendors now provide a wide range of SISs, such as basic software and control solutions for functional safety, **Safety Integrity Level** (**SIL**) evaluation, and alarm management. Certain vendors specialize in configurable field devices, such as safety controllers and sensors, as well as solutions for monitoring multiple sites.

The following overview depicts the leading industrial automation vendors and their associated SIS products:

Company	SIS Products
ABB	AC800M
	HI Safeguard
	Plantguard
	Triguard
Emerson Process Management	DeltaV SIS
GE	GMR System
	PAC8000 SafetyNet
HIMA	41q
	H51q
	HIMatrix
	HIMax
Honeywell	FSC DMR
	FSC QMR
Invensys (Schneider Electric)	Tricon
	Trident
Rockwell Automation	ControlLogix
	GuardPLC
	GuardLogix
	Trusted
RTP	RTP 2500
	RTP 3000
Siemens	SIMATIC S7-400FH
	QUADLOG
Yokogawa	ProSafe-RS
	ProSafe-SLS
	ProSafe-PLC

Table 2.2 – Major SIS vendors and their products

> **Important note**
> The preceding products are only for informational purposes and do not imply or suggest any kind of endorsement of these vendors.

Today, the ongoing emphasis placed on safety and compliance, combined with a heightened awareness of the need for industrial safety protocols, has led to the widespread adoption of safety protocols and processes within process facilities. This surge in demand has spurred the global SIS market to experience robust growth due to advancing trends in **Artificial Intelligence (AI)**, **Industrial Internet of Things (IIoT)**, and **Advanced Process Control (APC)**. Yet, sizeable concerns around both product quality and production efficiency continue to be a further catalyst to the adoption of SISs for risk mitigation.

Let's now look at the drivers for protecting SISs from safety and cybersecurity perspectives.

The need for protecting SISs

Industrial Control Systems (ICSs) and SISs are fundamental to ensuring the physical processes in process facilities, plants, power generation, or other critical infrastructure segments that do not exceed predetermined limits, as well as protecting valuable equipment, averting environmental harm, and keeping personnel safe.

If these precautions are not taken, the consequences can be incredibly serious. Imagine a plane traveling at an altitude of 10,000 feet and the safety systems fail; if the pilot is unable to retain control of the aircraft, the outcome could be catastrophic; the failure of these systems to do their job effectively will likely lead to both the destruction of the plane and to the loss of the passengers on board. Therefore, it is imperative that safety systems are robustly maintained and kept in full working order so that any potential accidents and catastrophes can be avoided.

SISs help ensure the proper functioning of **safety instrumented functions (SIFs)** and guard against the potential risks posed by hazardous events that may disrupt the process.

Safety Instrumented Function (SIF)

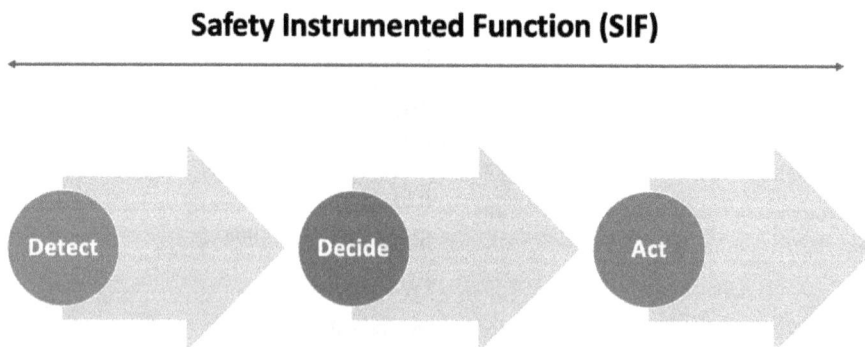

Figure 2.2 – Purpose of SIF

The ability of these SIF elements (sensors, logic solver, and final elements) to detect an imminent hazard, make decisions, and execute necessary actions – detect, decide, and act – ensures a safe state; however, the reliability of this will depend on the performance of the SIL assigned to the function.

The SIF purpose can be summarized in three key steps:

- **Detect**: The sensors must possess the capability to specifically detect the potential hazard and relay this crucial information to the logic solver in order to ensure a timely response and mitigate the risks involved

- **Decide**: A logic solver needs to possess the intelligence to identify when a hazardous situation has arisen, triggering the final element to react

- **Act**: The final element must have the capability to either guarantee that the process is back to a safe state or to provide the required countermeasures to mitigate any potential risks.

Some SIF examples include initiating a shutdown in hazardous processes, opening an emergency pressure relief valve to regulate tank overflow through an on/off switch, adding coolant to stop an exothermic runaway event, activating an automatic shutdown when no manual control is present or available to terminate a feed valve to prevent tank overflow, deploying fire suppressants, and signaling an evacuation alarm.

If one of the preceding safety functions fails, gets manipulated, or is intentionally or unintentionally interfered with, the following can be impacted directly or indirectly, depending on various conditions:

- **Safety**: SIS attacks can lead to a dangerous situation in which safety is compromised, potentially resulting in the injury or death of personnel and damage to property.

- **People**: An attack on an SIS can cause delays in operations, leading to personnel having to work extra hours or miss important deadlines. It may also result in a loss of trust from both stakeholders and customers. It can also lead to loss of life or injuries.

- **Financial**: An attack on an SIS can lead to financial losses due to the need to replace or repair damaged equipment, as well as the cost of any fines or legal action that needs to be taken as a result of the attack.

- **Asset**: An attack on an SIS can lead to the loss or damage of assets, including equipment, data, and intellectual property.

- **Environment**: An attack on an SIS can lead to environmental damage, such as pollution or contamination, if the attacked system provides protection against such hazards.

- **Reputation**: An attack on an SIS can lead to a loss of trust from stakeholders and customers, as well as negative media coverage, all of which can damage an organization's reputation.

Process industries have already witnessed several catastrophic safety incidents that changed the industry and, in the process, led to the introduction of new rules and standards. Some of the most highly documented incidents include the following:

- 1974: Flixborough (UK) vapor cloud explosion

- 1976: Seveso (Italy) TCDD cloud

- 1984: Bhopal (India) MIC cloud (US company)

- 1988: Piper Alpha (UK) oil platform fire

Additionally, in March 2005, the explosion of the AMOCO refinery resulted in 15 deaths, 180 injuries, and 8 people in critical condition. As well as the tragic loss of human life, the total cost to the company has since been estimated to exceed USD 2 billion.

The bottom line is that the potential consequences of the items on the preceding list are incredibly serious, with several elements rapidly evolving over the last decade. While certain developments are common across all ICS environments, we want to focus now on the current trends and risks faced by the modern ICS security environment in general and SISs in particular. So, what new trends and challenges are currently impacting the industry?

ICS cyber risk trends and evolution

The current state of ICS cyber risk is continuously evolving as malicious actors develop new and unexpected attack strategies to exploit these systems. While most industrial and manufacturing organizations and companies have, by now, adopted basic safety and security measures for their ICSs, which are effective in terms of identifying the risk level in their existing systems, firms are increasingly faced with the need to implement and maintain more complex, comprehensive, and dynamic security solutions that are capable of addressing the ever-changing cyber risks and threat landscape.

These are just some of the current trends that we are witnessing at present:

- **Digitalization**: There are visibility, architecture, and governance issues across organizations due to changes in business models and the adoption of new technologies and considerations of digital transformation initiatives, including IT/OT convergence.

- **Awareness**: The increase in awareness and wider acceptance within the industry about emerging threats targeting ICS environments – the systems that automate and control industrial operations – is an increasingly urgent challenge.

- **Geopolitical tension**: There is a shift in the global geopolitical atmosphere and the impact on defense strategies on critical infrastructures that have become a primary target for various adversaries.

- **Supply chain**: This is a deep and complex challenge for industries when it comes to managing third-party suppliers.

- **Cyber warfare**: The weaponization of ICSs, focusing on physical cyberwarfare and the deployment of a variety of tactics, such as disinformation, deception, denial of service, and distortion.

- **Organized cybercrime**: Cybercriminals are consequently turning their attention to infiltrating sectors where security barriers are less mature. Industrial sectors are becoming a growing target and lucrative target, as we have witnessed with the ransomware campaigns.

- **Shortage of skills**: For example, the scarcity of experienced and qualified staff after the COVID-19 pandemic, especially within the cybersecurity space.

- **Remote access**: The popularity of remote operations within ICS environments has grown in response to the former COVID-19 restrictions, resulting in organizations and vendors increasing their preventive maintenance deployments.

As for the challenges facing SISs, they include the following:

- SISs have flaws that can be attributed to poor testing, coding practices, and engineering:

  - The majority of components are insecure by design; as such, embedded systems are likely to be vulnerable.

  - The security of SISs depends on their software configuration and secure coding practices.

  - The need for unified vulnerability and risk management while also compensating any controls as part of a Defense-in-Depth strategy throughout the system lifecycle.

- Hardware or software failure and human errors are very relevant factors, according to various studies and statistics. The following figure depicts the results from an investigation of failures, including ICSs, that was carried out by the Health and Safety Executive (HSE) in the UK:

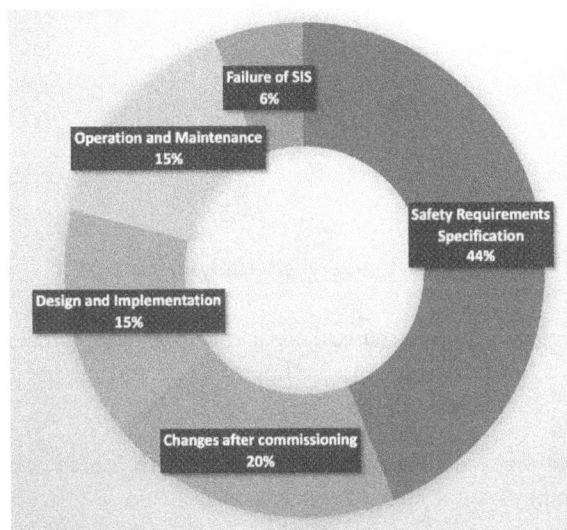

Figure 2.3 – Control system root cause of failure (HSE)

- Legacy patterns, which are inherently insecure by design, continue to exist and are unlikely to disappear in the near future.

- Unless serious security risks are disclosed, suppliers and asset owners may be unlikely to take appropriate actions to protect their systems.

- Operating systems, such as Windows NT, XP, and Windows 7, that are no longer formally supported by the vendor may still be in use in ICS environments.

Let's explore the intersection of safety and cybersecurity closely in the next section.

The intersection of safety and cybersecurity

The need to interconnect safety and cybersecurity is becoming ever more apparent. With greater convergence, engineers must merge the knowledge from these two overlapping areas in order to keep the potential related risks at a minimum. As safety is becoming increasingly dependent on cybersecurity, experts in both disciplines must understand how these two areas overlap and interact with one another in order to ensure effective system safety; there is now an increasing necessity to bridge the gap between safety and cybersecurity practitioners.

A heightened understanding of this intersection is invaluable when it comes to mitigating risk and promoting cybersecurity best practices while also guarding against possible safety-related threats. Additionally, each discipline must be aware of the vulnerabilities that are created within the system when safety and cybersecurity intersect. Given this, it is imperative that the safety and cybersecurity communities come together closely in collaboration; this is the only way to create efficient means for the successful integration of these two vital topics.

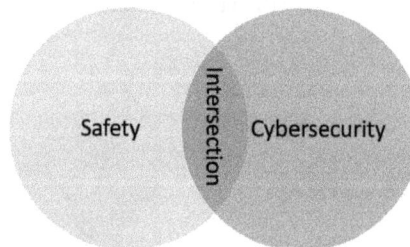

Figure 2.4 – Safety and cybersecurity intersection

Beyond the critical role of safety critical systems, there are many challenges that should be taken into careful consideration when it comes to the intersection of safety and cybersecurity. Some of these challenges include but are not limited to the following:

- **Terminology and cultures**: Confusion can arise due to differences in safety and cybersecurity terminology and culture.

- **Goals and objectives**: Inequity in objectives, operations, and outlooks can potentially result in the waste or inefficient use of time and resources.

- **Risk schemes**: There can be a lack of cohesion of a cybersecurity risk reduction scheme (in comparison to safety) since each component may have differing objectives and expectations. This might be a dilemma for quantitative and qualitative approaches to risk assessment.

- **Different pace**: There is a potential for conflict between the need for safety (static) and the ever-changing complexities of cybersecurity (dynamic) best practices, making it difficult to balance the two.

- **Undefined scope**: There can be deficient demarcation points and poor system boundaries regarding security that protects systems from malicious people and safety that protects people from a system malfunction.

- **Different philosophies**: There can be significant differences regarding risk paradigms and philosophies, which must be considered and addressed.

- **Obsolescence**: ICSs face significant obsolescence and legacy installation challenges compared to IT systems.

While the intersection between safety and cybersecurity appears to be small at first glance, there are positive signals and opportunities on the horizon, which will help strengthen collaboration and be driven by increased safety and cybersecurity awareness and the wider acceptance and enforcement of essential safety and cybersecurity standards.

For instance, here are some examples of initiatives through which companies can benefit from the intersection of safety and cybersecurity:

- Embedding cybersecurity within the process safety lifecycle

- Integrating cybersecurity risk assessment with **Health, Safety, Security, and Environment (HSSE)** risk assessment

- Sharing and leveraging information from internal and external networks

Now, let's learn more about the wider threat landscape and the key players with the potential to have an impact on the safety and reliability of operations.

ICS threat landscape

The ICS cyber threat landscape has become increasingly complex and unpredictable in recent times. As technology continues to advance at a rapid pace, more sophisticated attack vectors and methods are continually being developed to target ICSs and networks. Among the notable threats in recent years are ransomware, **Advanced Persistent Threats (APTs)**, malicious insiders, supply chain threats, and smart components, all of which can lead to ICS cyber compromise.

Increasingly, these threats are leveraging new attack surfaces provided by advanced technologies, such as the **Industrial Internet of Things (IIoT)**. IIoT devices and systems have proven to be very attractive targets for adversaries due to their low-security postures, lack of internal monitoring capabilities, and the difficulty of remote patching. This pushes cybersecurity teams to continually seek new and innovative ways to protect their ICS environments from malicious actors.

At the same time, the use of malware has become much more commonplace, with cybercriminals using ransomware to lock up critical systems and critical infrastructure, and financially motivated hackers are utilizing malware with the intention of stealing confidential, commercially sensitive data from companies. Malware is significantly evolving, becoming more sophisticated and advanced with each passing year, and, as such, it is imperative that organizations take steps to improve their security posture and take a proactive stance when it comes to defending their essential ICS networks.

Industrial assets are becoming more and more lucrative and attractive for various adversaries, and many are currently the target of various actors. In this context, we have grouped the sources of threat under the following categories:

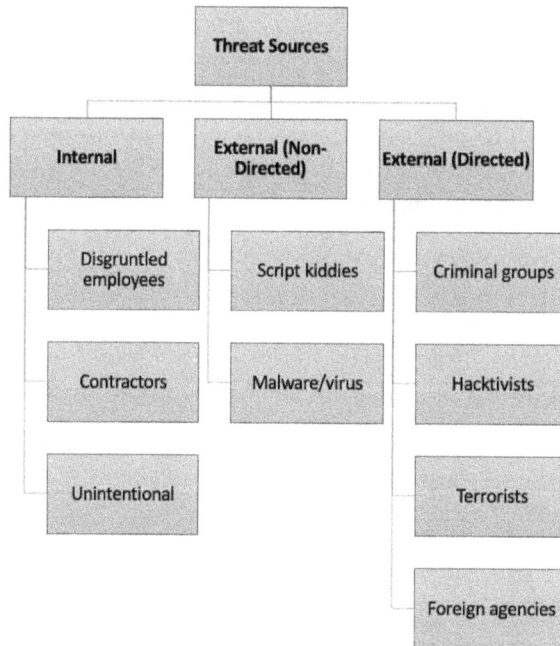

Figure 2.5 – Example of threat sources

The preceding diagram highlights examples of key adversaries that can threaten ICS environments. Each category consists of various threat actors with specific tactics and motivations:

- **Internal threats** involve those with access to an organization's systems, such as employees, contractors, and third parties. This can include operators, engineers, managers, technicians, contractors, and vendors, as well as network and system administrators who may be disgruntled or act maliciously. This insider risk can manifest as data theft, the manipulation of production data, leaking confidential information, or other malicious activities.

- **External (non-directed)** cyber threat sources refer to the practice of exploiting weak users and systems for malicious activities. This practice is usually carried out by "script kiddies" via a range of malicious software, including viruses, worms, keyloggers, and Trojan horses, in order to gain access to sensitive information or disrupt operations.

- **External (directed)** threats usually involve groups or individuals that deliberately target systems or networks for malicious purposes, such as stealing data, disrupting services, or extorting money. External threat actors can include criminal groups, hacktivists, terrorists, and foreign agencies. Criminal groups will typically focus on financially motivated attacks, such as fraud or the theft of data for financial gains, whereas hacktivists tend to be motivated by a social or political cause and are chiefly focused on disrupting business services or deleting data. Terrorists often use cyberattacks to achieve their goals and may use malware to infiltrate, control, or exfiltrate data from networks. Foreign agencies generally use cyber espionage to attain intelligence for economic and/or political gain or to paralyze critical infrastructures, all within the wider context of conflict.

When it comes to the execution of offensive activities, threat actors have a broad attack surface, a wide arsenal of tools, and known and unknown (zero-day) **Tactics, Techniques, and Procedures (TTPs)** with which to craft and launch attacks that often go unnoticed. The means and attributes of these threats include the following:

- Ransomware
- Social engineering
- Information gathering
- Infiltration
- Supplier compromise
- Access to a network or system
- Data exfiltration or system changes
- Disabling a safety system by preventing bypass removal
- Malicious logic solver firmware updates
- Software exploitation

> **Important note**
>
> Zero-day exploits are programs (payloads) that target known vulnerabilities in a system. These attacks are typically employed prior to the release of any security patches, meaning that the system remains unprotected and vulnerable. In the context of ICS networks, these attacks can be very difficult to respond to, as they rely on exploiting existing weaknesses in a system's security architecture. They are often difficult to trace or detect and can be used to gain (remote) unauthorized access, disrupt services, or steal data. Many known ICS security incidents (i.e., Stuxnet, Triton, and BlackEnergy) were found to be the result of zero-day deployment. There are even marketplaces and the dark web that sell such services.

The preceding TTPs may sound implausible for SIS and ICS environments, yet numerous examples have proven that these tactics – which may seem basic and comparatively unsophisticated – can be employed successfully with the help of teams from engineering, IT, and other disciplines all working together to achieve a collective goal.

The following screenshot is of the MITRE ATTACK Navigator, which can used for the analysis of threat actor TTPs:

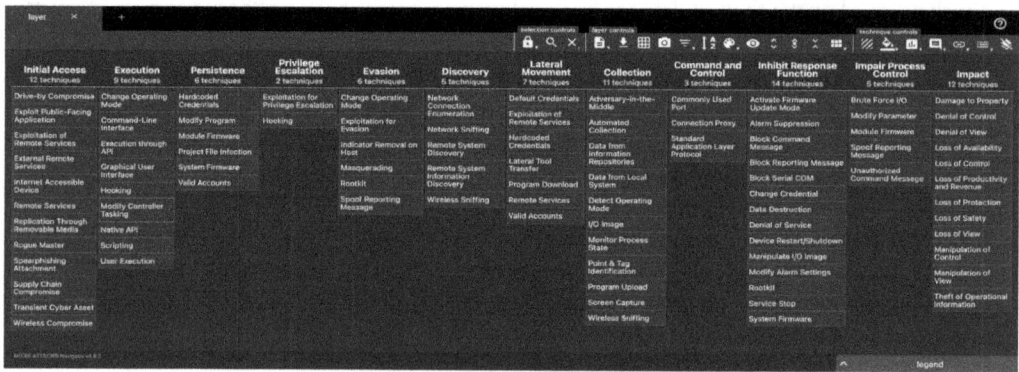

Figure 2.6 – MITRE ATTACK Navigator

(Source: `https://mitre-attack.github.io/attack-navigator/`)

In order to track the modus operandi of these various threat actors, the MITRE Corporation developed the MITRE ICS Matrix framework to help organizations assess and improve the security of their ICSs. The MITRE ICS Matrix provides a structured approach for analyzing and prioritizing security controls based on industry best practices and real-world threats.

It consists of two dimensions: adversary behavior and security functions. Adversary behavior includes different types of threat actors, such as nation states, hacktivists, and insiders. Security functions represent various security controls and capabilities that organizations can implement to protect their ICSs.

For more information about MITRE ICS Matrix, please visit the following link: `https://attack.mitre.org/versions/v13/matrices/ics/`.

At the time of writing of this book, there remain a significant number of assets that are exposed directly to the internet, including safety critical systems (see the example illustrated in the following screenshot). The exposure of control systems to the internet poses significant risks to their integrity, availability, and confidentiality. An unprotected connection can make them susceptible to unauthorized access, manipulation, or even total disruption.

The potential for targeted cyberattacks on these systems should not be overlooked; adversaries with malicious intent may seek to exploit vulnerabilities or weaknesses in the exposed assets to carry out physical cyberattacks. The ramifications of a successful attack on safety critical systems could be devastating, impacting not only the organization but also the broader community and infrastructure.

There are various search tools and platforms that can help you check whether your organization's ICS assets are exposed directly to the internet. The most popular ones are the following:

SHODAN.io: SHODAN is a search engine that provides public access to information about internet-connected devices and systems, allowing users to effectively search for and identify various devices, including servers, refrigerators, cameras, and industrial control systems, among others, that may be exposed and vulnerable to cyberattacks

Censys.io: Censys is a search engine that allows users to discover information about devices and networks on the internet, including details about open ports, SSL certificates, and other vulnerabilities that may be present, enabling organizations and individuals to assess potential security risks and take proactive measures to protect their systems and data

Fofa.info: Fofa is a search engine that helps individuals and organizations find specific data and information related to internet-connected devices, services, databases, and more, allowing for targeted searches and analysis to support various cybersecurity and risk assessment activities

The following screenshot shows the results from an online search for exposed safety controllers from Rockwell Automation:

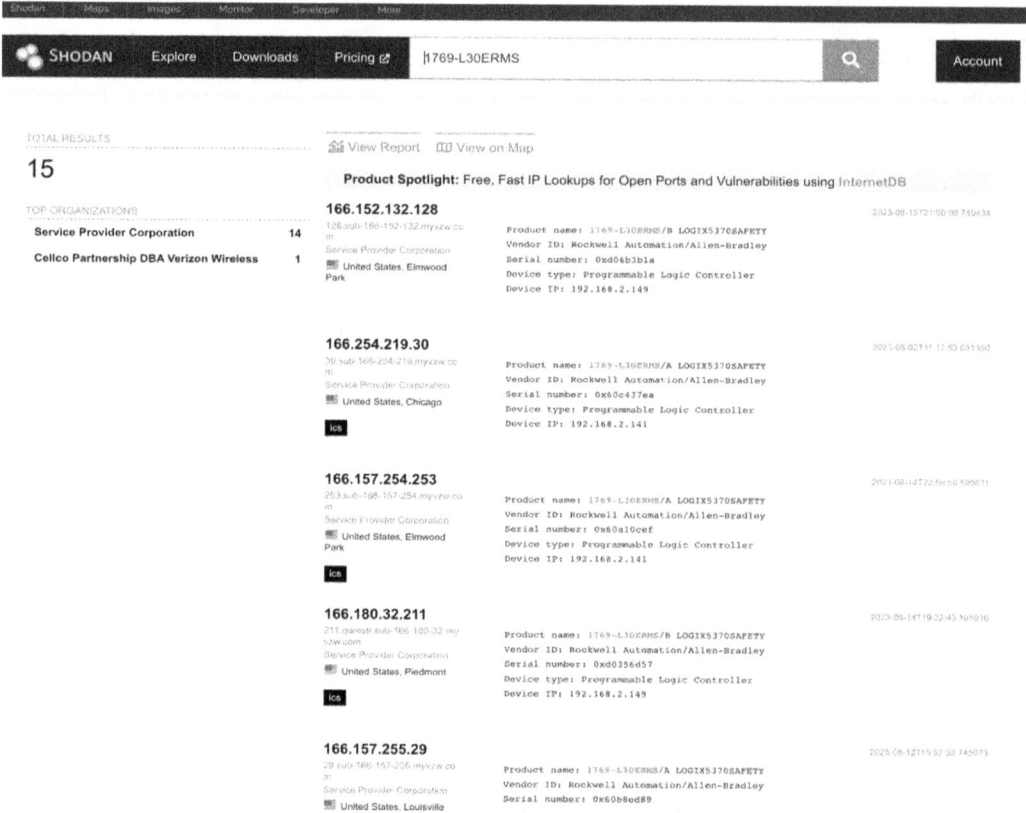

Figure 2.7 – Shodan search for exposed safety controllers

Exercise

Here is an exercise to become familiar with the search engines mentioned:

Task 1: Log in to one of the preceding search engines (SHODAN, Censys, or Fofa) and try to find ICS assets that are exposed to the internet and belong to your organization by using your organization's public IP addresses

Task 2: Use known industrial protocols such as OPC and Modbus

Task 3: Now, by using known product names, try to find any safety controllers

Tip: Consider using a filter cheat sheet and be aware of honeypots!

Let's now review some examples of cyber incidents that were made against an ICS environment and, in particular, against SISs.

ICS cybersecurity incidents and lessons learned

More than 500 known ICS cyber incidents have occurred worldwide to date (at the time of writing). Most of these were unintentional, but some were malicious attacks.

Their impact ranges from trivial to major outages regarding equipment damage and the total loss of production:

- **Colonial Pipeline**: In May 2021, the Colonial Pipeline Company was the victim of a successful ransomware attack that caused a system-wide shutdown. The closure and diminished oil transport capabilities had a severe impact on society because the Colonial Pipeline provides a substantial portion of the gasoline consumption on the East Coast of the United States. The full shutdown lasted for six days as the company addressed the attack (source: `https://www.energy.gov/ceser/colonial-pipeline-cyber-incident`).

- **Norsk Hydro**: In March 2019, the Norwegian global supplier of aluminum, Norse Hydro, experienced a significant security incident that negatively impacted the operation of its ICS networks. This resulted in several plants having to operate processing facilities manually, with an estimated cost of USD 50 million. It is conjectured that the control of safety functions, which are normally automated, was either diminished or not present; thus, manual processes were required to shut off the plant if an unsafe event arose. However, no safety concerns were reported (source: `https://www.hydro.com/en/media/news/2019/operational-and-market-update-first-quarter-2019-alunorte-and-cyber-attack-lower-overall-production-levels/`).

- **TRITON**: In 2017, TRITON (also known as TRISIS) malware was discovered; this program is believed to have been developed by a state-sponsored hacking group and was engineered to alter the firmware of SIS controllers. This poses a great risk, as the worst-case scenario could replicate the 1986 Chernobyl disaster, where a nuclear power plant's malfunctioning SISs caused an inability to efficiently govern temperatures and pressure levels. Whether a disaster is averted or not, the SIS malware may result in system shutdowns, thus rendering operations helpless until the SIS is operative again, leading to considerable business loss. Furthermore, with the hackers' intentions of securing additional targets, the risk of similar incidents increases (source: `https://www.cisa.gov/sites/default/files/documents/MAR-17-352-01%20HatMan—Safety%20System%20Targeted%20Malware_S508C.pdf`).

- **German steel mill**: In 2014, a disruptive cyberattack was launched against a German steel mill. The attack caused a malfunction in a blast furnace and other equipment, resulting in the total shutdown of production (source: `https://www.bsi.bund.de/SharedDocs/Downloads/DE/BSI/Publikationen/Lageberichte/Lagebericht2014.pdf?__blob=publicationFile&v=2`).

- **Shamoon**: Known as W32.DistTrack, Shamoon is a computer virus that was discovered in 2012 and primarily targets 32-bit NT kernel versions of Microsoft Windows. Its ability to spread between computers on the same network gives it high destructive potential, demonstrated by the substantial costs associated with recovery after an attack. This was exemplified in the attack on Saudi Aramco and RasGas, where "Cutting Sword of Justice" claimed responsibility for the deployed virus, incapacitating 30,000 workstations and making it one of the biggest hacks in history (source: `https://www.enisa.europa.eu/publications/info-notes/shamoon-campaigns-with-disttrack`).

- **Maroochy wastewater**: In March 2000, Maroochy Shire Council in Queensland, Australia, faced big challenges with its newly implemented wastewater system after recognizing disruptions in communication signals sent via radio frequency (RF). The system's pump functions were impaired, and the alarms intended to notify engineers of possible problems were not working as planned. Upon conducting an investigation, a monitoring engineer found that an intruder was purposely causing these disruptions, which resulted in the release of millions of gallons of untreated sewage into nearby rivers and local parks. In response, the water utility recruited a team of private investigators who were able to identify the perpetrator and notify the proper authorities. It was determined that the offender was a former employee of a SCADA software supplier who had been rejected for a local government role – a prime example of an insider attack (source: `https://www.osti.gov/servlets/purl/1505628`).

- **Stuxnet**: Stuxnet is an advanced piece of malware that was discovered and identified in 2010. It was primarily designed to target industrial control systems, in particular, **Supervisory Control and Data Acquisition (SCADA)** systems – including safety systems – and is believed to have been used to target centrifuges in Iran's nuclear program. Outwardly, the attack exploits a vulnerability in Microsoft Windows by using a stolen digital certificate for authentication as well as a zero-day flaw in the Siemens PLCs. A series of other strategies have since been used to avoid detection, including polymorphism and characteristics typically used by legitimate software projects (source: `https://csis-website-prod.s3.amazonaws.com/s3fs-public/2023-07/230703_Significant_Cyber_Incidents.pdf?VersionId=ez6_A71Z2QycWxK16FjD1nLx_CWR2o5z`).

In addition to these incidents, several alerts and advisories have been released concerning vulnerabilities exposed in several ICS products – including SISs – in order to help protect organizations from any potential security threats. More information can be found at trustworthy sources, such as the Cybersecurity and Infrastructure Security Agency CISA (`https://www.cisa.gov/news-events/cybersecurity-advisories?search_api_fulltext=&sort_by=field_vendor`) or ICS Strive (`https://icsstrive.com/`).

The analysis of these incidents has exposed the ultimately evolved technical capabilities of threat actors, as well as their willingness to inflict physical damage. In particular, Stuxnet and TRITON demonstrated that cyber activity can have a serious, real-world impact.

When viewed from this perspective, the ability to detect, resist, and recover from cyberattacks is of paramount importance to critical infrastructure owners and operators. Additionally, it has become increasingly evident that knowledgeable adversaries can leverage various techniques to infiltrate and hijack such systems, and information-stealing malware is often deployed to begin their course of action.

Furthermore, it has become clear that cybercriminal groups and nation states are actively developing capabilities to attack critical infrastructure. The range of malware and ransomware now used by various threat actors indicates that they have unlimited capacity to craft and launch sophisticated cyberattacks against their targets. In addition, attackers are showing a real willingness and capability to launch malicious, destructive, and disruptive attacks on those infrastructures depending on the mission objectives.

> **Important note**
> Major incidents in the industry are now leading to significant changes in attitudes among asset owners, operators, and suppliers. From calculating the related costs, they are now considering fees from unforeseen high-impact events, including insurance and/or the revocation of a license to operate.

Summary

This chapter examined the history and evolution of SISs, exploring the major trends, as well as the biggest challenges currently faced. We explored examples of known incidents and their consequences and provided an introductory understanding of the key concepts and terminology.

The next chapter will delve much more deeply into SIS design and architecture.

Further reading

- ENISA Threat Landscape 2023:

    - `https://www.enisa.europa.eu/topics/cyber-threats/threats-and-trends`

    - `https://www.csa.gov.sg/alerts-advisories/Advisories/2022/ad-2022-004`

- ARC Advisory Group:

    `https://www.arcweb.com/blog/who-accountable-safety-systems-cybersecurity`

- ICS Strive:

    `https://icsstrive.com/`

- Code of Practice: Cyber Security and Safety:

  ```
  https://electrical.theiet.org/guidance-codes-of-practice/
  publications-by-category/cyber-security/code-of-practice-cyber-
  security-and-safety/#:~:text=This%20Code%20of%20Practice%20
  is,a%20threat%20of%20cyber%20attack.
  ```

- ABB 800xA:

  ```
  https://new.abb.com/control-systems/safety-systems/system-800xa-
  high-integrity
  ```

- Emerson DeltaV SIS:

  ```
  https://www.emerson.com/en-us/automation/control-and-safety-
  systems/safety-instrumented-systems-sis/deltav-safety-
  instrumented-system
  ```

- HIMA:

  ```
  https://www.hima.com/en/industries-solutions
  ```

- Honeywell Safety Manager:

  ```
  https://process.honeywell.com/us/en/products/control-and-
  supervisory-systems/safety-systems/safety-manager
  ```

- Rockwell:

  ```
  https://www.rockwellautomation.com/en-us/capabilities/process-
  solutions/process-safety-systems/safety-instrumented-systems.html
  ```

- RTP Corporation:

  ```
  http://rtpcorp.com
  ```

- Schneider Electric Triconex:

  ```
  https://www.se.com/il/en/product-range/63681-triconex-safety-
  systems/?subNodeId=12366861404en_IL
  ```

- Siemens:

  ```
  https://www.siemens.com/global/en/products/automation/systems.
  html
  ```

- Yokogawa ProSafe:

  `https://www.yokogawa.com/us/solutions/products-and-services/control/control-and-safety-system/safety-instrumented-systems-sis/`

- Chemical Safety Board (CSB):

  `https://www.csb.gov/investigations/completed-investigations/?Type=2`

- HSE Safety Incidents:

  `https://www.hse.gov.uk/eci/incidents.htm`

- HSE. *Out of control Why control systems go wrong and how to prevent failure*:

  `https://www.hse.gov.uk/pubns/priced/hsg238.pdf`

- SHODAN:

  `https://www.shodan.io/`

- Censys:

  `https://censys.com/data-and-search/`

- Fofa:

  `https://en.fofa.info/`

- The Industry IoT Consortium:

  `https://www.iiconsortium.org/`

3

SIS Security Design and Architecture

In the age of the **Internet of Everything (IoE)**, **Artificial Intelligence (AI)**, and the fast-evolving digital space, security awareness is becoming an increasingly important part of everyday life. With organizations storing a growing amount of sensitive data and personal information, it is no surprise that people are becoming more mindful and vigilant when it comes to potential security threats that can affect their privacy and financial records. Yet most of the public is not aware of the security issues that face critical infrastructure. One of the most relevant and important areas today is understanding the security of **Industrial Control Systems (ICSs)** – systems that monitor, control, and protect operations and production environments – and, in particular, **Safety Instrumented Systems (SISs)**. These systems are often tasked with managing the safety of mission critical assets and, as such, need to be designed to guarantee the utmost security.

As many SIS deployments include legacy designs that are not updated or properly maintained, they lack the security features to adequately protect these systems from threats. Furthermore, inefficiencies in the architectural designs of these systems inhibit the development of the architectural components, which further exacerbates the issue of safety among ICSs. On the other hand, this has caused some to believe that the traditional method of safety and isolation – such as an *air gap* or physical security – is sufficient protection for an SIS; with digital protection, there will always be the possibility of bypass or compromise.

This chapter will provide a detailed analysis of the design and architecture of ICS networks, with a specific focus on SISs. It will explain their components and security protocols and will provide popular SIS implementation examples in order to offer a wider context and illustrate their safety-oriented structures. By the end of this chapter, you should have a greater understanding of the security challenges related to SIS design and architecture, as well as the best practices for both.

We will cover the following topics in this chapter:

- Understanding **Defense in Depth (DiD)** and **Layers of Protection Analysis (LOPA)** principles
- ICS security design and architecture
- ICS key components for safety functions
- SIS secure architecture
- Example SIS reference architectures
- Safety network protocols
- ICS proprietary protocols

Understanding DiD and LOPA principles

To ensure the safety and security of ICS environments, it is essential to prioritize and diversify protection philosophies with a focus on well-proven principles. Ideally, deploying both DiD and LOPA are essential steps to take to achieve safety and cybersecurity goals. These strategies strive for the same end result of reducing risks by integrating various methods of protection and safeguarding levels. Deploying both DiD and LOPA allows for comprehensive protection of the system from both a design and operational level with multilayered methods that are tailored to fit various conditions. These strategies are highly complementary and work to reduce the consequence of potential events ensuring comprehensive protection of safety and the security of ICS environments.

Let's now explore, in this section, the specifications of these strategies in detail.

The DiD strategy

Safeguarding industrial networks calls for a comprehensive defense strategy. In light of the steady increase in cyber-attacks targeting process industries, relying on a single layer of protection for safety-critical assets is now insufficient. DiD is generally viewed as an effective strategy in ensuring the security of an ICS network against potential cyber-attacks, which consists of multiple independent security barriers that complement each other. When implemented properly, it can increase the probability of successfully repelling a malicious attack and ensure attackers can be met with increased levels of difficulty when trying to gain unauthorized entry to the network.

The DiD strategy is based on physical and logical tactics that can provide deterrence to safeguard assets, as listed in the following diagram:

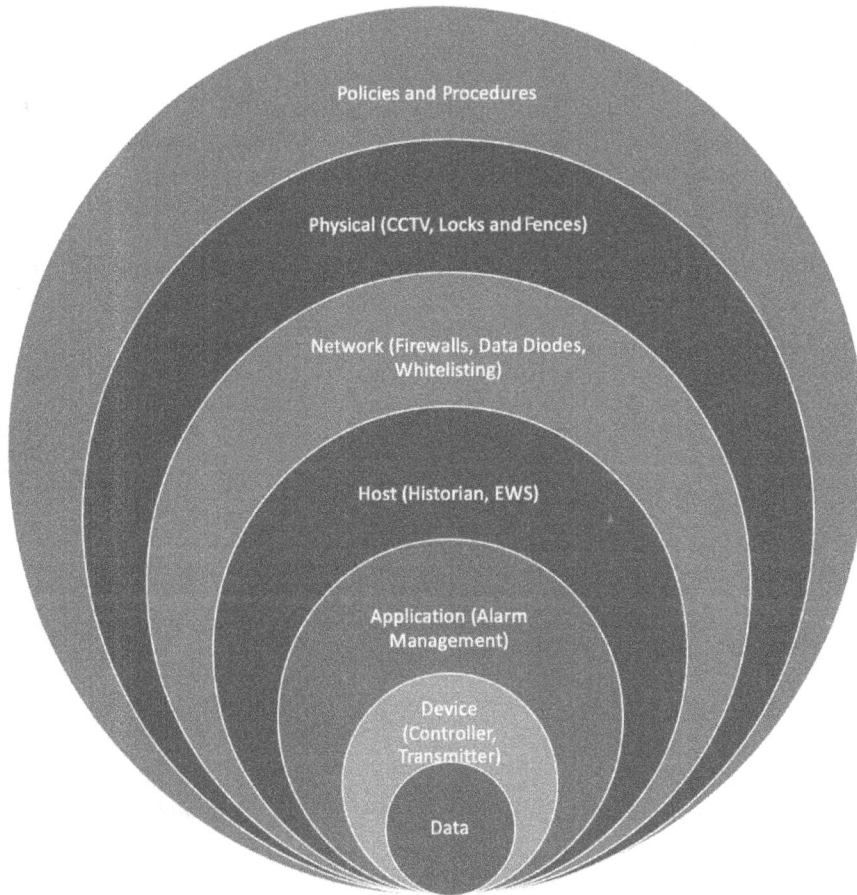

Figure 3.1 – DiD strategy

You can consider DiD as multiple layers of locks that a threat actor would be required to unlock or bypass in order to gain access to the ICS network and safety-related applications or devices. The sophistication of these barriers increases once we move deeper into the network and interface with safety critical systems.

A comprehensive DiD approach encompasses multiple layers of security to effectively mitigate cyber risks at various levels. This includes a strong governance foundation with well-defined security policies and procedures, as well as physical safeguards to prevent unauthorized access and protect assets from any potential sabotage, theft, or vandalism. Additionally, network security measures are implemented to safeguard the network perimeter against external threats and ensure proper segmentation between IT and ICS domains, as well as within ICS networks between **Basic Process Control Systems (BPCSs)** and SIS and physical process levels. Vertical and horizontal protection strategies are employed to ensure comprehensive protection, with endpoint and application security being enforced at the host level, while field devices and process data are constantly safeguarded.

The following table includes an example of these barriers and a high-level overview of their functions:

Barrier	Description
Policies and procedures	Defines the requirements related to security in controls, with a focus on process, data, control, and safety functions in the ICS context
Physical security	Ensures that the organization's facilities, people, and critical assets such as SISs are protected from internal and external threats
Network	Requires solid architecture and technologies (that is, firewalls, data diodes, proxies, whitelisting) to detect and mitigate any potential attacks
Host	Is deployed to secure endpoint devices (desktops, servers, and so on) from malicious threats such as malware, ransomware, social engineering, and so on
Application	Involves configuration and hardening of applications to protect against attacks
Device	Requires solid security baseline and hardening to reduce attack surface
Data	Takes measures to keep the integrity and confidentiality of process data secure from tampering or hacks

Table 3.1 – Examples of security barriers

As threat actors become increasingly sophisticated and technology advancements move forward at a rapid pace, coupled with explosive numbers of breaches, this strategy can no longer rely solely on protection but requires additional pillars. If prevention fails, your security controls should be in a position to instantly detect the breach and trigger a timely and adequate response to successfully contain the breach. These pillars are based on the **National Institute of Standards and Technology (NIST)** framework and consist of the following:

- Preparation
- Prevention
- Detection
- Response

To effectively practice DiD, organizations need to find out which methods will work best for their context and maturity. This includes selecting suitable technologies such as firewalls, data diodes, anomaly detection, and elements to apply to their strategy.

Chapter 5, *Securing Safety Instrumented Systems*, will cover full details of ICS security controls and the NIST framework.

> **Important note**
>
> The selection of security technologies for ICS production or a safety critical mission environment is highly crucial. Organizations, ICS security professionals, and suppliers should be confident about product capabilities and the potential negative impact that can cause damage or harm to live systems. This is not an area of experimentation as the potential consequences of an incident or near miss could have severe ramifications.

The LOPA strategy

As a similar strategy deployed to process safety, LOPA states that no single layer of defense can absolutely eradicate risk since a certain level of risk always exists when it comes to protective measures such as **Safety Instrumented Functions (SIFs)**, pressure relief valves, or even physical containment barriers such as dikes. (Dikes are constructed to control or regulate water flow, prevent flooding, or protect land from erosion).

LOPA is a simplified methodology used to calculate how each layer contributes to risk reduction, and the use of SISs plays a major role in this to achieve a higher level of safety. The following diagram shows how LOPA protection barriers and mitigation actions are manifested to prevent top events:

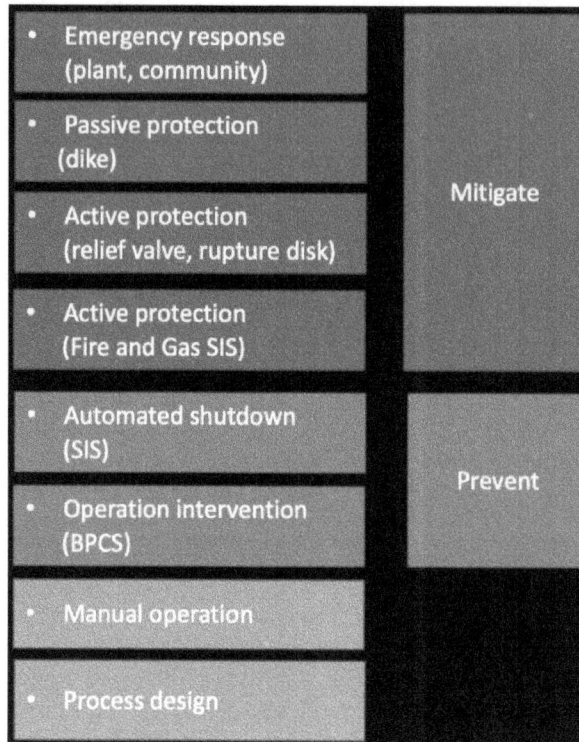

Figure 3.2 – Safety protection layers

For each risk, LOPA recommends a specific layer of protection designed to mitigate the risk of that consequence. The following table highlights the seven layers of protection that are recommended by LOPA to address hazardous process operations:

Barrier	Role
Emergency response	Depending on the type of consequence, an emergency response may be required. Establishing a system to respond to an emergency can include activities such as establishing an **Emergency Response Plan (ERP)**, providing emergency response equipment such as fire extinguishers, and training personnel in emergency response techniques.
Passive protection	This layer is used to limit or reduce the probability and severity of the hazardous event. Examples of passive protection can include physical barriers to limit access, engineering controls such as emergency vents or pressure relief valves, and redundant systems.

Barrier	Role
Active protection	This layer helps to detect if a hazardous event has occurred and provide an automated response. These can include **Fire Detection Systems (FDSs)**, smoke and heat detectors, **Leak Detection Systems (LDSs)**, and pressure sensors.
Automated shutdown (SIS)	This layer detects when an abnormal process condition has occurred and automatically shuts down the affected system. In fact, the SIS functionality is present in both the prevention and mitigation layers. The **Emergency Shutdown (ESD)** functions are categorized under the prevention layer, while the fire and gas SIS functions fall under the mitigation layer, activating systems such as deluge systems, fire systems, and warning systems.
Operation intervention (BPCS)	This layer monitors process parameters and allows operators to control the process and prevent hazardous conditions.
Manual operation	This layer provides human operators with the ability to recognize process conditions and take manual control through the use of manual operating procedures, alarms, and direct action.
Process design	This layer identifies how the process should be designed so as to minimize risks and liabilities associated with hazardous operations. This can include redundant systems, safe handling techniques, and control systems.

Table 3.2 – Layers of protection

In the next section, we will explore the ICS design and architecture according to industry best practices.

ICS security design and architecture

Due to the immense size and complexity of ICS, which encompasses multiple components in distinct geographic areas running multiple processes simultaneously, systems are often split into multiple operational zones. Each area requires distinctive technical specifications, resulting in numerous distinct models being built for effective management. These legacy systems were not initially created with security in mind; as such, they are markedly susceptible to cyber threats if proper security strategies are not put in place.

The implementation of a DiD strategy and the incorporation of the Purdue model could greatly contribute to a secure architectural design.

The Purdue model

The Purdue model was first introduced as a guide to illustrate the movement of data in **Computer-Integrated Manufacturing (CIM)**, an advanced system for fabrication where computers take control of the whole manufacturing process. It later developed into an open standard for the formation of ICS/**Operational Technology (OT)** frameworks that enable appropriate security – achieved through the implementation of different layers and an orderly progression of data flow and interconnectivity between key ICS components. The Purdue model is now widely used within the ICS community and is certainly a good entry point for a secure architecture and DiD strategy.

In accordance with the Purdue model, ICS networks are generally separated into four distinct zones, as depicted in the following diagram:

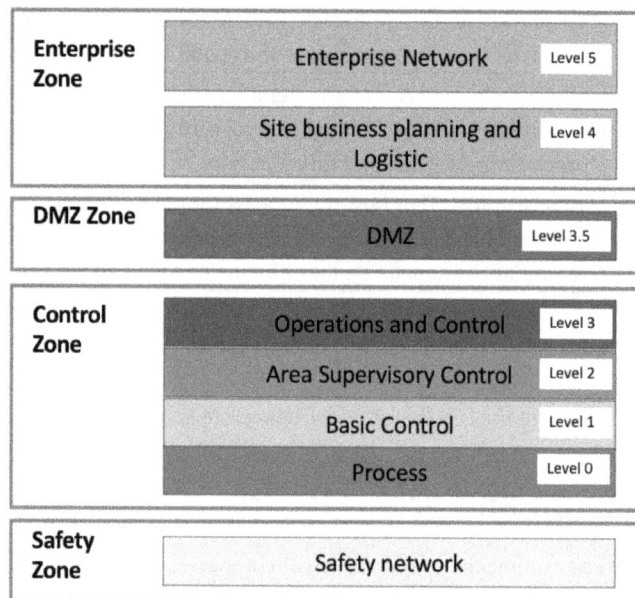

Figure 3.3 – ISA99 zones and levels

We will analyze the four zones, accompanied by their respective levels (ranging from Level 0 to Level 5) of the Purdue model:

- **Enterprise zone**: Operates independently from ICS, but crucially relies on data from ICS networks to inform business decisions. This is where business applications reside, such as **Enterprise Resource Planning (ERP)**, **System Applications and Products in Data Processing (SAP)**, or others in which supply chain activities are conducted. Nowadays, the majority of these systems are hosted in the cloud.

- **DMZ zone**: Serving as a bridge between IT and OT systems, this includes a designated jump host for secure remote access and shared services such as file transfer, patching, and anti-virus measures for ICSs.

- **Control zone**: Encompasses devices and systems for automated control of processes, production lines, or **Distributed Control Systems (DCSs)**. In contemporary ICS solutions, Levels 1 and 0 are often integrated.

- **Safety zone**: This is where the safety function resides, especially safety controllers.

The segregation of ICS networks into seven distinct levels, as prescribed by the Purdue model, is a frequently used approach for creating an adequately defended and secure infrastructure. When correctly implemented, this division of ICS networks enables the establishment of a robust environment. The following diagram illustrates the main components and levels of the Purdue model:

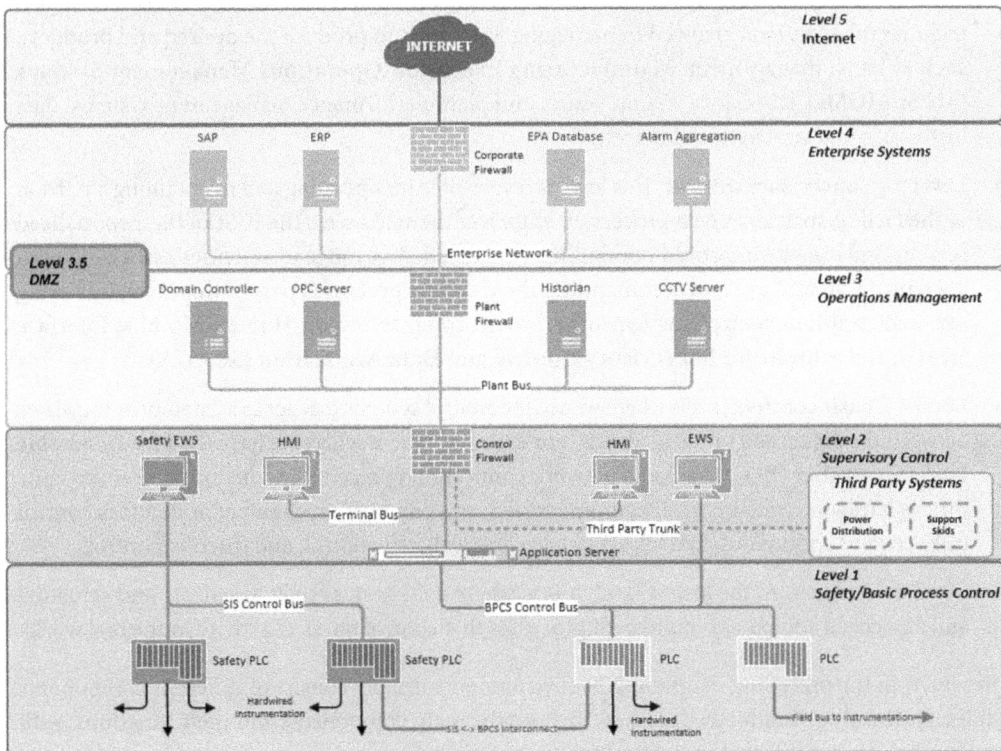

Figure 3.4 – Purdue components and levels

Let's investigate how different systems interact with one another by taking a deep dive into these levels:

- **Levels 4 and 5 – IT enterprise network**: The top two levels represent the IT network, which nowadays is extended to the cloud for business applications. The IT network includes systems that communicate with ICS networks and/or allow access to the other levels via the **Demilitarized Zone (DMZ)**.

- **Level 3.5 – OT/IT DMZ**: The ICS DMZ acts as a secure border between the ICS and the IT networks. Information can be shared without exposing the ICS components to IT networks or the internet.

- **Level 3 – Site operations**: In larger environments, you could have multiple areas under supervisory control (see *Level 2*). When multiple systems of systems exist within the entire site, all roll up to be monitored and controlled as a single entity. This level can also be responsible for collecting all relevant process data within the site to provide to the business in the IT network. It also includes the functions involved in managing workflows to produce the desired end products, such as batch management, **Manufacturing Execution/Operations Management Systems (MESs/MOMs)**, laboratory, maintenance, and plant performance management systems, data historians, and related middleware).

- **Level 2 – Supervisory control**: This level is responsible for operating and maintaining a portion of the facility, such as a given process, to empower them to assess the ICSs of their specialized regions and execute improved operations when needed. A range of activities associated with keeping a watchful eye and commanding the concrete process is part of this level, including oversight, real-time controls implemented through computerization, **Human-Machine Interface (HMI)**, and software for **Supervisory Control and Data Acquisition (SCADA)**.

- **Level 1 – Basic control**: This is where we see the control systems that act as a gateway or translator between the digital and physical worlds. For example, here is where we have the **Programmable Logic Controller (PLC)** that has a network connection to receive instructions and at the same time is hardwired into physical equipment such as a combustion chamber. It includes control tasks such as continuous control, sequence control, batch control, and discrete control.

- **Level 0 – Process**: At the lowest level, this is where field devices such as sensors and actuators and associated equipment implement processes that control physical activity in the real world.

As presented in the preceding diagram, ICS environments usually consist of different components that interact directly or indirectly. Let's now explore the main components and their functions, with a focus on the components that are critical for safety functions.

ICS key components for safety functions

Components of ICSs comprise controllers, software applications, field devices, and communication devices. This can extend in certain environments to include mission-specific components such as telecom systems, subsea, **Building Management Systems** (**BMSs**), and **Industrial IoT** (**IIoT**). This section explains the main types of ICSs, along with other component types and their conventional purpose:

- **Historian (Process Historical Archiver (PHA), historian, or logger)**: The historian acts as a repository for historical data gathered from intelligent devices within a control system. It can operate independently or in conjunction with an operator workstation. Usually, the PHA functions through a Windows-based PC. It offers a streamlined method for managing and retrieving transferred data and is compatible with various industrial protocols such as Modbus and **Open Platform Communications** (**OPC**), allowing for direct connections to HMIs, PLCs, and **Remote Terminal Units** (**RTUs**) for data retrieval. Unlike a standard IT database system, it doesn't support referential integrity in tables and has the unique capability to ingest and store a large amount of data quickly.

Next is an illustrated depiction of a historian's overview of alarms and events for a power plant:

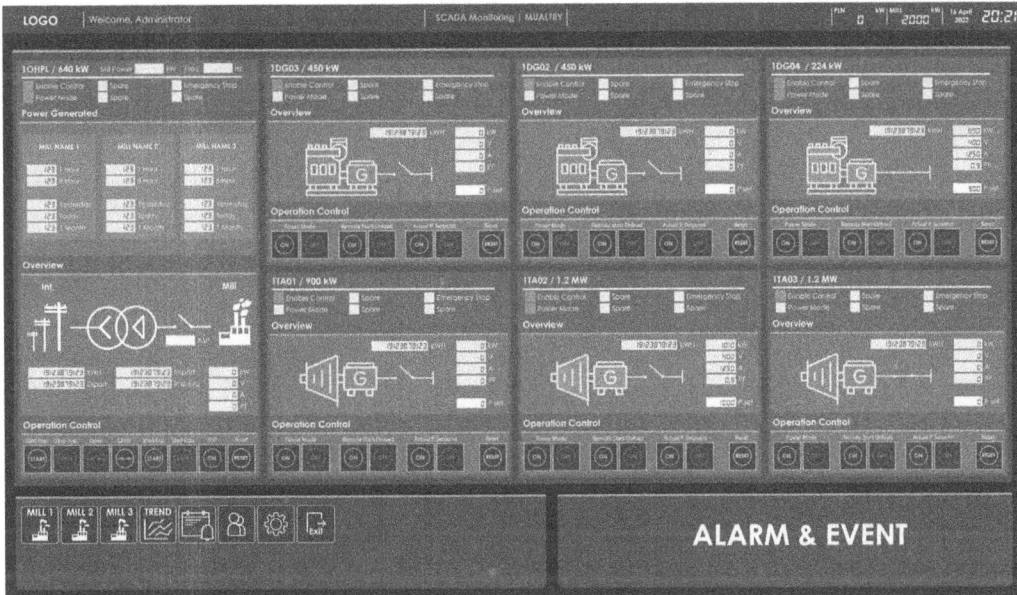

Figure 3.5 – Example of a historian's overview

- **HMI**: An HMI enables plant operators to monitor process values, alarms, and data trends and manipulate automation processes from an external platform. An HMI usually facilitates communication between a user and a machine, program, or system. Although the technical definition extends to any screen used for device interaction, it is more commonly used in industrial contexts. HMIs provide real-time data and allow for the control and configuration of machinery through a **Graphical User Interface (GUI)**:

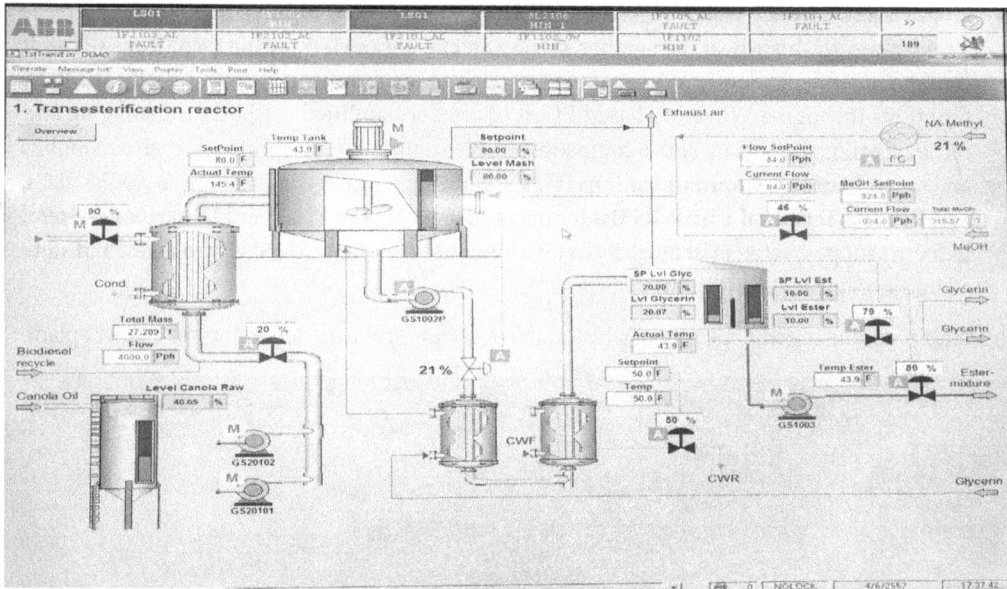

Figure 3.6 – Example of an HMI

- **Engineering Workstation (EWS)**: The use of EWSs within process facilities allows for efficient management, configuration, and oversight of production processes. Typically, these workstations consist of desktop computers equipped with an operating system and specialized ICS software that is capable of facilitating both operational and engineering tasks. If the capabilities and policies of the end user's system allow, a single workstation can effectively serve as both an **Operations Workstation (OWS)** and an EWS:

Figure 3.7 – EWS

- **OWS**: Typically, a Windows-based PC that serves as the main interface for the operator and offers a range of features such as color graphics, faceplates, alarms, logging, trends, and diagnostics. The EWS also contains an OWS that is specifically utilized for testing and resolving issues.

- **Application workstation (AWS) or app station**: A hybrid system that often includes database management, historical data analysis, online display, specialist configuration, and system monitoring.

- **Safety EWS**: The safety EWS is usually dedicated to safety functions, not processing any tasks related to basic process control as an EWS does. As several vendors and organizations prefer to integrate safety and control functions, there is usually one EWS that supports both safety and control duties.

- **PLCs**: PLCs can be used for managing and monitoring any part of the process and are designed to permit expansion via removable **Input and Output (I/O)** devices, communication modules, and memory units. Standard PLCs enable complex logic, arithmetic, and data manipulation operations to be achieved through a choice of distinct programming languages – ladder logic, structured entering, and function block diagrams. For extra safety features, a standard PLC may be further supplemented by external safety apparatus or structures:

Figure 3.8 – PLC

- **Safety PLC**: A safety PLC is a special type of PLC that is designed to perform safety-related functions in critical applications. It has redundant hardware, software, and communication components that ensure high reliability, **high availability (HA)**, and **fault tolerance (FT)**. A safety PLC can detect and react to faults, errors, or hazards in the system and initiate appropriate actions, such as shutting down the process, activating alarms, or switching to a safe state. A safety PLC follows strict standards and regulations, such as IEC *61508* and ISO *13849*, that define the **Safety Integrity Level (SIL)** and the **Performance Level (PL)** of the system:

Figure 3.9 – Safety PLC

- **Instrument Asset Management System (IAMS)**: The IAMS is a specialized system designed to facilitate the tracking and managing of critical ICS. It assists technicians and engineers by providing a comprehensive environment for monitoring performance, safety, and reliability. The IAMS provides an integrated network of functions and hardware components that continuously monitor asset health, record information pertaining to instrument performance, detect faulty instrument elements, log event record information, and provide quick response or recovery to pre-defined conditions and criteria.

  With an IAMS, personnel can analyze data gathered from the control system, provide trend detection capabilities, and gain insights to optimize and improve the control system. An IAMS can also provide personnel with improved visibility of performance status data, enabling a greater focus on proactive actions that will minimize the financial, operational, and public safety implications of instrument system problems.

- **Field devices**: Present in various industrial systems, field devices are technological equipment responsible for translating physical changes or actions into signals that are recognizable by the control system; they include sensors, transducers, actuators, and other monitoring equipment. Such devices are connected directly to a controller (including a PLC, an RTU, or an **Intelligent Electronic Device (IED)**) through a digital or analog I/O module or an industrial protocol such as Modbus or Profibus.

- **Converters or communication gateways**: A converter can be used to facilitate communication between two systems that utilize different protocols and/or transmission mediums. *Figure 3.10* highlights this in further detail, illustrating that the converter must transform data sent from the sending system to match the protocol and transmission medium of the host system. A notable example of this is the conversion of messages sent using Modbus on Serial (*RS-232/RS-485*) to a format that can be recognized by OPC messages on Ethernet:

Figure 3.10 – Protocol converter

- **Sequence of Events (SoE)**: One of the features commonly found in SIS implementations is the ability to record an SoE. This function is able to record precise event data from digital inputs with a 1 ms resolution through **Digital Input** (**DI**) modules. The system's behavior can be accurately recorded due to the generation of events with each scan cycle. The storage capacity for events is significant, with up to 15,000 events being able to be stored in each controller. Furthermore, 500 pre-trip events and 1,000 post-trip events can also be stored when trip events are specified.

Over recent years, IEC *62443*, NIST, and various initiatives from the ICS security community have introduced new concepts for securing architecture through network segmentation and visibility. In the next section, we will explore the value that these concepts can add to protecting ICS networks.

ICS zoning and conduits

ICS zones and conduits play a vital role in ICS cybersecurity. OT is broadly regarded as a set of technologies used to observe and govern physical devices and operations, such as production systems in ICSs. A security zone makes up a logical assembly of physical, informational, and application resources that share the same security requirements and is an essential component of a DiD approach.

Conduits then serve as the physical connections between these security zones, not only to restrict access to important systems but also to restrict malicious traffic and protect the network traffic's consistency. Firewall policies, access controls, and other security steps – when combined with conduits – can create the boundaries of a controlled network that can aid in shielding legacy ICS assets.

Security networks deploy conduits to regulate access to important systems, while simultaneously blocking out any potentially malicious traffic and ensuring process data remains uncompromised. By maintaining these conduits and combined security resources such as firewall protocols, access denials, and other related measures, a control system network can be enforced that will safeguard important ICS assets.

The following diagram shows examples (presented in red circles) of zones and conduits:

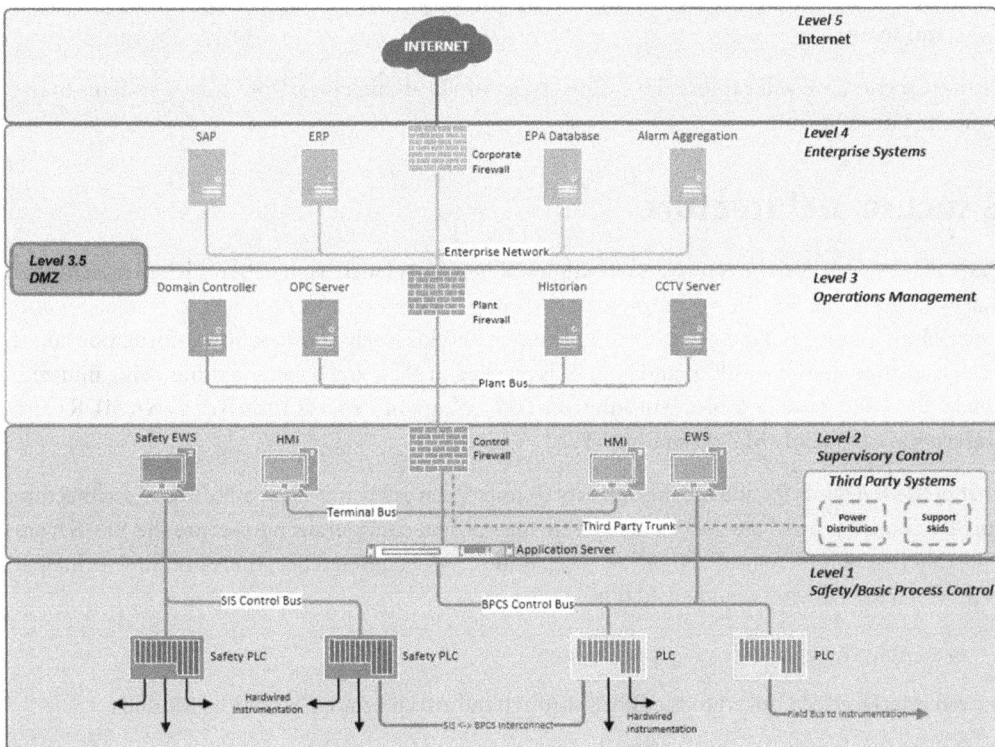

Figure 3.11 – Zones and conduits

Conduits can be deployed in various forms and include several components, such as the following:

- Firewalls
- Data diodes
- Routers with **Access-control Lists (ACLs)**
- Proxies
- RS-232/422/485 serial cable

When modeling zones and conduits, it is important for those involved to bear the following in mind:

- Zones can have sub-zones, but conduits cannot have sub-conduits
- Cyber assets (hosts) within a zone require the use of either one or more conduits for communication, yet a conduit cannot pass through more than one zone
- A conduit can be used by two or more zones to communicate between them

The traffic that flows through the conduits is called channels, and this can include industrial protocols (including OPC, Modbus, PROFINET…and so on) or enterprise protocols (such as HTTP, FTP, SSH, Telnet…and so on).

In the next section, we will explore the various types of SIS architecture, from legacy systems to the new modern deployments.

SIS secure architecture

We have discussed ICS architecture and examples of incidents (such as the 2017 TRISIS attack) that primarily targeted the SIS. This prompted asset owners, operators, and suppliers to reconsider and reevaluate how safety functions can be better protected and to closely consider how information could be safely exchanged between BPCSs and SISs. This resulted in a few architecture options being initiated, driven by the **User Association of Automation Technology in Process Industries (NAMUR)** and the **International Society of Automation (ISA)**.

Neither of the standards for either cybersecurity or functional safety prescribe a necessary architecture for an SIS. Thus, it is up to the user to determine the optimal configuration to secure the BPCS from the SIS, in terms of both logical and physical separation. As a consequence, there are three principal ways in which SIS networks can be laid out:

- A completely separated, air-gapped system
- An interfaced SIS linked to the BPCS through industrial protocols, such as Modbus
- An SIS that is joined with the BPCS yet sufficiently isolated and follows a strict security baseline

No matter which SIS setup and design is adopted, a security posture should be determined beforehand during the design and implementation phases in order to maximize security. While certain individuals argue that an isolated, air-gapped system is safest, the security requirements and corresponding risks of a business can vary dramatically, making each architecture a matter of preference.

SIS architecture types include the following:

- **Isolated SISs**: It is evident that the creation of an *air gap* between the core SIS functions and the BPCS provides a form of protection that is effective at shielding the system from cyber intrusions. However, it should not be assumed to be a foolproof security measure. This is due to both the fact that external access to the system is often needed and the considerable amount of time and effort that it may take to manage two distinct DiD architectures:

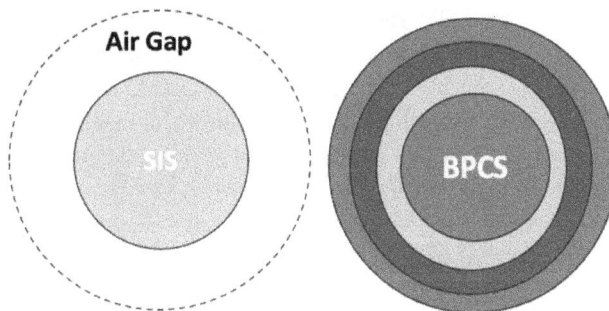

Figure 3.12 – Air-gapped SIS architecture

- **Interfaced SISs**: Interfaced systems are similar to separated systems in that safety-based processes are isolated from other ones. However, updated links constructed with mandatory open protocols are used to build the correlation between the BPCS and the SIS. Mechanisms such as security hardware, firewalls, and software are employed to limit the collaboration between the BPCS and SIS. Having the core SIS dissociated from peripheral utilities, these linked systems have a superficial approach to compliance where the architectural or organizational separation does not guarantee true security or compliance. To obtain full protection, multiple overlapping security steps must be taken on all connected sources, which may require extra efforts to be made for monitoring. It is ultimately the responsibility of the user to make sure that the tie between the BPCS and SIS is handled safely in order to avoid any possible risk:

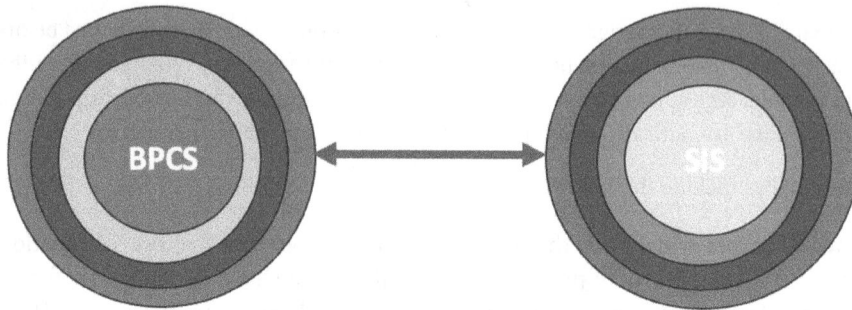

Figure 3.13 – Interfaced SIS architecture

- **Integrated SISs**: An integrated SIS provides an alternative to separate systems for engineering. In this approach, the SIS and BPCS are connected but maintained as both a logically and physically separate entity. Security measures lack inconsistency from manual endeavors, due to collaboration between the two using a proprietary protocol that operates distinctively with in-built cybersecurity. Significantly, this enhanced safety protocol heightens DiD protection, all the while minimizing the time and effort required to attend to the updating and maintenance of security layers, vis-à-vis creating a more fortified security barrier. In addition, an integrated SIS includes specific additional security protocols to protect the core SIS baseline:

Figure 3.14 – Integrated SIS architecture

After discussing each type of architecture in detail and exploring the security risks associated with them, we will determine which type of architecture is most appropriate. Identifying this is dependent on specific types of mission and business requirements. This brings us to one of the important rules when it comes to considering SIS and BPCS connection architecture: a safety connection architecture must support the specific business requirements and associated business/mission risk. In other words, the architecture needs to be tailored to the specific organizational needs.

We can now look at the general security characteristics of the different connection architectures and what to consider when selecting an architecture for specific requirements. The following table summarizes some of the advantages and limitations of the three implementations:

Architecture	Benefits	Limitations
Isolated	• Robust physical protection • Data transfer is conducted by restricting the transmission of information through analog hardwiring among the I/O cards of the safety controller and the I/O cards of the process controller • In case of an attack on the HMI, IAMS, or EWS, there are no interfaces to SIS	• Challenging for operation, especially for the plant startup process • Rely on local investigation and support, which is time-critical for process safety • Deviate from operational excellence goals and latest trends for preventive maintenance and analytics • Challenging and an increased risk in the case of a crisis (for example, COVID-19 – people are not willing to go to sites or hazardous areas)
Interfaced (Serial)	• Well grounded • Industry standard within the engineering community	• Costly for commissioning and testing • Serial connections are not secure proof as they can be subject to cyber-attacks • More physical security is required
Integrated	• Security relies on the configuration and engineering security skills • Comes with additional security features • Cost-effective	• Increases attack surface massively • Cascading effect in case of failure and huge dependencies between systems • Known single point of attack (interfaces)

Table 3.3 – SIS architecture-type benefits and limitations

The proficiency of the SIS in terms of reliability, security, and maintenance depends greatly on the version of the SIS. Earlier versions were typically air-gapped and isolated, but recent developments have seen SISs integrating increasingly in the age of convergence.

For instance, legacy SISs contain no active support or product development from vendors. The tools for engineering are linked directly to a serial or other type of communication port on the SIS. The connection to the HMI employs a proprietary gateway or serial connection that cannot be routed to the process control network, and field devices exclusively utilize analog and discrete signals.

As for the intermediate SIS, these systems moved into a mature support phase with limited or no major product developments from the vendor. Nowadays, the majority of engineering tools and utilities can be accessed through a control network that utilizes proprietary protocols. Furthermore, the gateway to the HMI operates as a terminal server or OPC server, or it can be directly connected through Modbus TCP to the process control network. It is important to note that most field devices still continue to rely on analog and discrete signals.

Integrated safety systems are supported and in an active product development cycle from the vendor. The engineering tools now reside on an S-EWS PC on the process control network, which may be on the same PC as the EWS, but with role separation for security. The engineering tool can be incorporated into both the database and HMI, allowing for an automatic population of the logic solver configuration. In current practice, the HMI utilizes a direct peer connection on the control network, previously relying on a proprietary protocol such as Vnet/IP or SIS-NET, but now implementing an open protocol such as Modbus TCP.

Furthermore, field devices use both analog and discrete signals for SIFs but are now taking on digital bus technologies such as wired Ethernet, PROFINET, and industrial wireless technologies such as WirelessHART and ISA100.11a for diagnostics and configuration – in turn connected to IAMSs that may reside on the control network. Additionally, **Partial Stroke Testing (PST)** for final shutdown elements is carried out through the process control network, which captures all PST outcomes. The integration of these systems provides a seamless and efficient safety process that is making the wider workplace safer.

In addition, several different communication protocols are available, including proprietary protocols via a proprietary gateway for non-routable communications, Modbus Serial (RS-232/485) via a proprietary gateway for non-routable communications, proprietary protocol via Ethernet (TCP/IP or UDP/IP) directly to the process control network, industry protocols such as Modbus TCP via Ethernet directly to the process control network, and Fieldbus communications directly to smart devices such as **Highway Addressable Remote Transducer (HART)**, ProfiSafe and Foundation Fieldbus, connected to multiplexers or other interfaces and, in turn, connected to the IAMS.

With all of the aforementioned characteristics, proper security measures should be always implemented to ensure the digital security of critical assets is addressed – for instance, updating software and hardware, installing firewalls, monitoring the network infrastructure, and using **Endpoint Detection and Response (EDR)** systems to protect against malicious activities.

Managing entry points

By limiting the available entry points – by means of firewalls, data diodes, and gateways – and introducing robust mitigations for any risks associated with them, it is possible to maximize the security of an SIS network to ensure its safe functioning. Careful consideration of DiD layers is an excellent starting point, but it is not enough to guarantee cyber safety. That said, the security of the SIS must take into consideration the robustness of the following key elements:

- HMI
- Safety EWS
- IAMS
- DCS

The following diagram represents an example of common SIS interfaces and entry points:

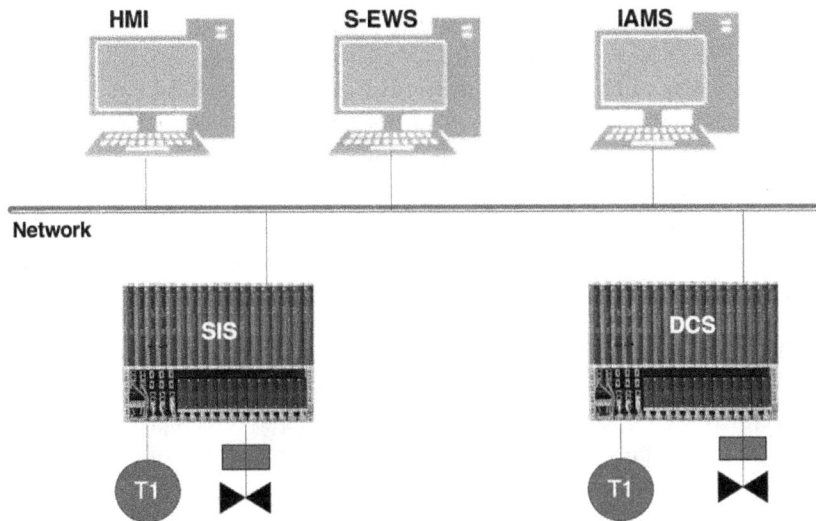

Figure 3.15 – SIS typical interfaces

SIS entry points and attack vectors will be covered in depth in *Chapter 4, Hacking Safety Instrumented Systems*.

In the next section, we will gain further insight into different architectures for major DCS/SIS deployments to date.

Example SIS reference architectures

Major vendors typically offer a total solution, which includes a DCS/SIS. These vendors each have a different architecture and control-safety philosophy. To give an idea of design and architecture in the context of cybersecurity, here are some examples of the most frequently used SIS implementations – as previously mentioned, deployment can range from fully integrated to interfaced or air-gapped safety-control connections.

Vendors typically have their own standards for naming systems within their architecture, though the engineering functions of these components remain the same. In addition, these environments are static, which means that naming conventions, IP addresses, and design all remain the same throughout the system lifecycle.

> **Important note**
>
> The intersection of suppliers' commercial interests colliding with safety control systems in the SIS community is a highly contested area of discussion. For some, the distinction between the two is clear – safety systems should remain completely physically separate from control systems and should not be purchased or developed from the same supplier. Others, however, suggest that modern technology and other techniques of risk reduction might be enough to ensure logical and functional separation. This idea is not a novel one, as evidenced by the neutrality of standards within the field concerning the integration of safety and control systems.
>
> Software solutions capable of detecting and avoiding faults have become more advanced, and certain types of integrated safety even now feature a degree of built-in variance from the development teams. For those who side with safeguarding independence, their cause is bolstered by the tangible evidence of previous missions being successfully accomplished. However, supporters of the integration of the two systems can demonstrate the cost and efficiency advantages of this choice with their own extensive base of case studies. Furthermore, third-party consultants are more apt to follow the more mathematically verified methods described by the standards than aligning with either side. In conclusion, while safety should not be compromised by higher productivity and less complexity, the risks of the operation can be appropriately managed in more than one way.

We have explored the SIS architecture and examples of deployments. Now, let's navigate our way through common safety network protocols used across different architectures.

Safety network protocols

Communications with the ICS environment thrived throughout the evolution of process control innovation as well as design and architecture. As we discussed in previous chapters, ICS environments were not primarily designed with security in mind; there remains a lot of proprietary information that is still not accessible to the public, including proprietary protocols.

International Standard IEC *61784-3* specifies the most favored functional safety networks, with special focus paid to the **Communication Profile Families** (**CPFs**) identified in the Fieldbus standardization ecosystem.

In this section, we will explore the most relevant protocols, with a focus on safety network protocols.

HART

The HART protocol is an international standard for communicating with intelligent field devices in process automation. It is a reliable, cost-effective technique for exchanging configuration and diagnostic information with sensors, transmitters, and other machine elements in an industrial process. The protocol is based on two-way digital communications over a 4-20mA analog signal path. It provides a cost-effective communication solution for monitoring a variety of industrial equipment, including smart sensors, transmitters, and valves. The protocol allows bidirectional digital communication with the device, making it possible to configure the device's settings, initiate diagnostics, and more accurately monitor the operation of the device.

Apart from the communication capabilities, the HART protocol comes with a number of advantages. Being a digital communication protocol, equipment manufacturers and plant engineers can benefit from these advantages by embracing the HART protocol. Advantages include remote monitoring, the availability of data for asset tracking, configuration and diagnostics, improved environmental performance, asset management, and automated data collection. Therefore, the usage of the HART protocol in automation systems is becoming increasingly widespread.

Modbus

Modbus is an industrial communications protocol that facilitates communication between intelligent electronic devices, such as microcontrollers, computers, PLCs, and other electronic devices, over an RS485 serial communication interface. It is used by the manufacturers of **Process Control Systems** (**PCSs**) and industrial automation equipment and is a popular industrial protocol used in a variety of applications, including motor control, data acquisition, industrial instrumentation, and machine automation. Designed to provide robustness and FT, Modbus accommodates up to 247 nodes on a single physical network over an RS485 communication link by utilizing a master-slave network configuration.

The Modbus protocol defines how information will be transmitted on the network, as well as the format and structure of the data. It also provides a mechanism for error checking, providing a measure of data integrity. Modbus is commonly used in industrial automation, process control, and data acquisition systems. With its cost-effective and efficient communication capabilities, Modbus is a popular choice for energy management systems and building or industrial automation systems. It is also widely used for process monitoring and for distributed control applications commonly found in the power, paper, cement, and chemical industries.

OPC

OPC is an industry standard protocol developed by the OPC Foundation. It is designed to provide secure and reliable communication between different systems, including those from different vendors. OPC is used in a wide range of industries, such as automotive, energy, pharmaceutical, aerospace, and many more. It facilitates communication between devices and systems and various parts of the manufacturing process. OPC allows for the exchange of data between systems and components in multiple languages and protocols by using a hub and spoke model for communication, which consists of one central hub and several remote devices. This allows them to communicate with each other in real time and enables rapid data exchange and fast response times, which is essential for efficient manufacturing processes.

OPC is based on open standards such as XML and **Simple Object Access Protocol** (**SOAP**) and is supported by most industry-leading software vendors. Furthermore, OPC is a secure protocol and provides access control that enforces security policies and encryption mechanisms. It also provides a tamper-proof log to ensure that data is not modified and is audit-secure. This ensures data integrity and provides secure data access to ensure system security. In a nutshell, OPC is an industry-standard protocol used to communicate between disparate systems based on **Object Linking and Embedding** (**OLE**) for process control. It allows for secure, reliable, and fast data exchange, giving manufacturers the ability to confidently process and manage data quickly and accurately. OPC is a critical tool that helps to ensure both efficient manufacturing processes and data security.

There are various OPC definitions to access process data, alarms, and historical data:

- **OPC Data Access (OPC DA)**
- **OPC Historical Data Access (OPC HDA)**
- **OPC Alarms and Events (OPC A&E)**
- **OPC Unified Architecture (OPC UA)**

SafeEthernet (HIMA)

SafeEthernet is a new communication protocol provided by HIMA, an international firm specializing in automation safety. Using contemporary technologies, it is designed to give users a secure, reliable, and cost-effective method of communication for automation installation in industrial settings. SafeEthernet is a safe, hardened backbone for automation communication, allowing the integration of multiple safety protocols such as PROFINET and Sercos. It is integrated with the **Active Network Layer** (**ANL**) for better performance and secure data transmission to securely connect to cloud-based systems. This allows users to access digital information such as operational and machine data from the cloud securely.

The protocol provides enhanced cybersecurity through authentication, claims-based security, Secure Boot, secure software updates, encryption, and private key storage. It also complies with standards such as IEC *62443-4-2* for industrial security processes and provides authentication through secure passwords and digital certificates. The SafeEthernet protocol ensures that the user's industrial automation system is secure and reliable while allowing secure access to cloud-based systems. This helps users to remain up to date on global standards, compete globally, and remain competitive in the ever-changing industrial landscape.

Vnet/IP (Yokogawa)

Vnet/IP is an industrial protocol from Yokogawa that facilitates Ethernet communication between devices. It is an open protocol that supports both wired and wireless connections between devices. This protocol helps to ensure that communication between devices is secure and reliable, as it encrypts the data and uses the existing local routing infrastructure for routing nodes.

Vnet/IP is a cost-effective solution for connecting industrial field devices as it does not require any additional hardware or software. It also provides real-time monitoring, remote access, and diagnostics. Yokogawa's Vnet/IP supports a wide range of industries such as manufacturing, process control, and energy management. It offers an extensive library of functions and a comprehensive set of services such as control, diagnostics, scheduling, alarm management, asset management, data logging, and historian services. This allows Vnet/IP to be used in a variety of applications and provides a flexible, secure, and reliable data exchange path. It is an ideal protocol for the modern industrial environment.

ProfiSafe

The ProfiSafe protocol is an industrial communications protocol used in industrial automation systems. It was developed by German firm Profinion to replace the traditional proprietary two-way communication systems typically used in this field. Its aim is to make different types of equipment from different vendors interoperate across diverse industrial automation networks. The main features of the protocol are its flexibility, scalability, reliability, and integrated safety concept. The protocol allows users to exchange both process data and safety data in the same packet. It also supports multiple fieldbuses, including Ethernet, Profibus, and **Controller Area Network** (**CAN**), allowing users to connect devices from different vendors.

Profisafe's safety concept covers the whole industrial process c from the application layer to the physical layer. This ensures that safety levels are maintained during communication between different equipment. Moreover, ProfiSafe meets international standards, such as ISO/IEC *61158* and ISO/IEC *61784*. Safety messages are structured and coded in a manner that allows for any type of automation system, including those of different vendors, to work together safely and efficiently. The protocol also ensures that data is safe from any potential threats, making it the best choice for demanding industrial automation networks.

Functional Safety Over EtherCAT (FSoE)

EtherCAT is a real-time Ethernet-based fieldbus system developed by Beckhoff Automation for measurement, control, and communication in industrial automation. It provides reliable and deterministic communication for control systems, reducing wiring costs and making automation easier to deploy. It is a cost-effective and reliable solution for a wide variety of industrial applications. EtherCAT has built-in features that provide safety over networks and is compliant with the following industry safety standards: IEC 62443, EN 50155, and IEEE 802.3.

The combination of EtherCAT with redundancy makes it a reliable safety protocol in terms of communications and data integrity. EtherCAT offers high-speed data rates of up to 100 Mbit/s, up to 255 nodes in a single cycle, multicast/broadcast capabilities, high deterministic performance, and a reliable safety protocol. This protocol is also compatible with a wide range of industrial devices, such as I/O, motion control, force control, and vision systems. As a result, EtherCAT is a reliable and cost-effective choice for industrial automation applications, offering robust and secure safety protocol features.

CIP Safety

The **Common Industrial Protocol Safety** (**CIP Safety**) network protocol, a product of the Open Group, is an open framework for industrial automation system security. It is designed to achieve greater security of control system networks and operational technology, providing asset owners and operators with a way to securely bring automation systems and the networks that connect them into the corporate information infrastructure. CIP Security seeks to prevent and reduce the risk of security threats to the corporate network, protecting data from unsafe access or unauthorized manipulation.

The protocol applies both proactive and reactive measures to keep operational technology secure, including mandated encryption over public networks, intrusion prevention measures, secure authentication of users, and secure identification of **Remote Access Points** (**RAPs**). CIP Security also provides secure decision support for analysis and the authorization of network connections, as well as assurance of secure data exchanges.

CC-Link Safety

CC-Link Safety is an industrial network protocol that ensures safe operations and secure data transmission for industrial automation systems. The protocol is designed for fieldbus communication between PLCs and other devices to protect equipment and personnel from potential hazards. It is specially tailored to meet the various conditions and complexities of safety requirements in the industrial sector.

The protocol uses an encrypted communication structure that reduces noise and crosstalk while also providing high-speed, low-cost operation. Additionally, it supports fault-tolerant communication structures that detect and correct errors in the network. Other features include support for redundant safety networks, host and fieldbus communication, simplified system management, and third-party supplier support. The protocol is certified at SIL 2 and SIL 3 safety integrity levels depending on the number of safety channels. CC-Link Safety allows for several safety-related functions such as safety advice, monitoring, response, and response decisions, and its safety-related features are capable of achieving high system safety objectives.

openSAFETY

The openSAFETY network protocol is a communication protocol that allows the sharing of real-time safety data among various safety devices and systems, such as radar, cameras, traffic signals, and other infrastructure. This protocol ensures that the exchange of data is secure, accurate, and private by using multiple layers of encryption, authentication, and access control.

 It also includes various other features, such as secure transmission of device-to-device messages – using "Black Channel" principles – encryption of information while in transit, and anti-counterfeiting measures. It is also designed to be extensible, allowing for new protocols to be added to the system as needed.

In the next section, we will learn about ICS proprietary protocols and their **Original Equipment Manufacturer (OEM)** or vendors.

ICS proprietary protocols

ICS proprietary protocols, unlike open-source protocols, are built and owned by a single organization. OEMs impose usage limitations, such as with a patent or their implementable trade secrets, which prevent their protocol from benefiting anyone outside their product reach. These protocols are tailored to just the organization's products and services and are not open for use by other developers or companies. On the flip side, open source offerings are adjustable and available to anyone, providing equipped bidirectional use.

The following table is an example of key proprietary protocols used within the ICS environment, in particular within process industries:

Vendor	Protocol
ABB	ABB 800xA DCS (IEC 61850 Manufacturing Message Specification (MMS) including ABB extension)
	Clock Network Communication Protocol (CNCP)
	Redundant Network Routing Protocol (RNRP)
	ABB Inter Application Communication (IAC)
	ABB Totalflow
Beckhoff	Automation Message Specification (AMS)/Automation Device Specification (ADS)
	TwinCAT
Emerson	DeltaV
	DeltaV Discovery
	Emerson OpenBSI/Bristol Standard Synchronous/Asynchronous Protocol (BSAP)
	Ovation DCS Advanced Digital Module Diagnostics (ADMD)
	Ovation DCS Data Processing Unit Status (DPUSTAT)
	Ovation DCS Single-Station Remote Processing Controller (SSRPC)
Emerson Fischer	Remote Operation Controllers (ROC)
GE	Bently Nevada (System 1/BN3500)
	ClassicSDI (MarkVle)
	Ethernet Global Data (EGD)
	GE Standard Messaging (GSM) (GE MarkVI and MarkVIe)
	InterSite
	System Data Interface (SDI) (MarkVle)
	Service Request Transport Protocol (SRTP) (GE)
	GE_CMP

Vendor	Protocol
Honeywell	ENAP
	Experion DCS Common Data Access (CDA)
	Experion DCS Fault Detection and Analysis (FDA)
	Honeywell Ethernet Universal Control Network (EUCN)
	Honeywell Discovery
Rockwell	Client Server Protocol 2 (CSP2)
	EtherNet/IP (ENIP)
	ENIP CIP (including Rockwell extension)
	ENIP CIP FW version 27 and above
	Modbus/TCP
Schneider Electric	Modbus TCP – Schneider Unity Extensions
	OASyS (Schneider Electric/Telvent)
	Schneider Triconex System Access Application (TSAA)
	Foxboro Evo
Schneider Electric/Invensys	Foxboro I/A
	Trident
	TriGP
	TriStation
	Control and Monitoring Platform (CAMP)
Siemens	PCS7
	PCS7 WinCC – Historian
	Profinet DCP
	Profinet I/O
	Profinet Realtime
	Siemens PHD
	Siemens S7
	Siemens S7 - Firmware and model extraction
	Siemens S7 – key state
	Siemens S7 Plus
	Siemens SICAM/WinCC

Vendor	Protocol
Yokogawa	Centum ODEQ (Centum/ProSafe DCS)
	HIS Equalize
	FA-M3
	Vnet/IP

Table 3.4 – Example of proprietary protocols (extracted from Microsoft Defender for IoT: https://learn.microsoft.com/en-us/azure/defender-for-iot/organizations/concept-supported-protocols

Next are packet captures of the main SISs:

- Yokogawa Vnet/IP:

```
Frame 1: 250 bytes on wire (2000 bits), 250 bytes captured (2000 bits) on interface unknown, id 0
    Section number: 1
    Interface id: 0 (unknown)
    Encapsulation type: Ethernet (1)
    Arrival Time: Aug  9, 2017 14:18:14.277191000 CEST
    UTC Arrival Time: Aug  9, 2017 12:18:14.277191000 UTC
    Epoch Arrival Time: 1502281094.277191000
    [Time shift for this packet: 0.000000000 seconds]
    [Time delta from previous captured frame: 0.000000000 seconds]
    [Time delta from previous displayed frame: 0.000000000 seconds]
    [Time since reference or first frame: 0.000000000 seconds]
    Frame Number: 1
    Frame Length: 250 bytes (2000 bits)
    Capture Length: 250 bytes (2000 bits)
    [Frame is marked: False]
    [Frame is ignored: False]
    [Protocols in frame: eth:ethertype:ip:udp:skype]
    [Coloring Rule Name: UDP]
    [Coloring Rule String: udp]
 Ethernet II, Src: YokogawaDigi_a0:c5:d1 (00:00:64:a0:c5:d1), Dst: IPv4mcast_40:18:01 (01:00:5e:40:18:01
    Destination: IPv4mcast_40:18:01 (01:00:5e:40:18:01)
    Source: YokogawaDigi_a0:c5:d1 (00:00:64:a0:c5:d1)
    Type: IPv4 (0x0800)
 Internet Protocol Version 4, Src: 192.168.129.128, Dst: 239.192.24.1
    0100 .... = Version: 4
    .... 0101 = Header Length: 20 bytes (5)
    Differentiated Services Field: 0x20 (DSCP: CS1, ECN: Not-ECT)
    Total Length: 236
    Identification: 0x9376 (37750)
    000. .... = Flags: 0x0
    ...0 0000 0000 0000 = Fragment Offset: 0
    Time to Live: 1
    Protocol: UDP (17)
    Header Checksum: 0xdb80 [validation disabled]
    [Header checksum status: Unverified]
    Source Address: 192.168.129.128
    Destination Address: 239.192.24.1
 User Datagram Protocol, Src Port: 32977, Dst Port: 9940
```

Figure 3.16 – Example of Vnet/IP packet capture

- Emerson DeltaV:

```
v Frame 1: 66 bytes on wire (528 bits), 66 bytes captured (528 bits) on interface \Device\NPF_{592491E5-D
    Section number: 1
  > Interface id: 1 (\Device\NPF_{592491E5-DF0F-4AB3-84F5-7D809437B93D})
    Encapsulation type: Ethernet (1)
    Arrival Time: Jan 25, 2018 11:11:25.862786000 CET
    UTC Arrival Time: Jan 25, 2018 10:11:25.862786000 UTC
    Epoch Arrival Time: 1516875085.862786000
    [Time shift for this packet: 0.000000000 seconds]
    [Time delta from previous captured frame: 0.000000000 seconds]
    [Time delta from previous displayed frame: 0.000000000 seconds]
    [Time since reference or first frame: 0.000000000 seconds]
    Frame Number: 1
    Frame Length: 66 bytes (528 bits)
    Capture Length: 66 bytes (528 bits)
    [Frame is marked: False]
    [Frame is ignored: False]
    [Protocols in frame: eth:ethertype:ip:tcp]
    [Coloring Rule Name: TCP SYN/FIN]
    [Coloring Rule String: tcp.flags & 0x02 || tcp.flags.fin == 1]
v Ethernet II, Src: VMware_be:8c:07 (00:0c:29:be:8c:07), Dst: FisherRosemo_26:72:5c (00:22:e5:26:72:5c)
  > Destination: FisherRosemo_26:72:5c (00:22:e5:26:72:5c)
  > Source: VMware_be:8c:07 (00:0c:29:be:8c:07)
    Type: IPv4 (0x0800)
v Internet Protocol Version 4, Src: 10.4.0.6, Dst: 10.4.0.14
    0100 .... = Version: 4
    .... 0101 = Header Length: 20 bytes (5)
  > Differentiated Services Field: 0x00 (DSCP: CS0, ECN: Not-ECT)
    Total Length: 52
    Identification: 0x4fd4 (20436)
  > 010. .... = Flags: 0x2, Don't fragment
    ...0 0000 0000 0000 = Fragment Offset: 0
    Time to Live: 128
    Protocol: TCP (6)
    Header Checksum: 0x0000 [validation disabled]
    [Header checksum status: Unverified]
    Source Address: 10.4.0.6
    Destination Address: 10.4.0.14
v Transmission Control Protocol, Src Port: 52004, Dst Port: 18519, Seq: 0, Len: 0
    Source Port: 52004
    Destination Port: 18519
    [Stream index: 0]
  > [Conversation completeness: Complete, WITH_DATA (31)]
    [TCP Segment Len: 0]
    Sequence Number: 0    (relative sequence number)
    Sequence Number (raw): 3081758563
    [Next Sequence Number: 1    (relative sequence number)]
    Acknowledgment Number: 0
    Acknowledgment number (raw): 0
```

Figure 3.17 – Example of DeltaV packet capture

- ABB 800xA:

```
Frame 1: 118 bytes on wire (944 bits), 118 bytes captured (944 bits)
    Encapsulation type: Ethernet (1)
    Arrival Time: Aug 13, 2018 08:26:01.194530000 CEST
    UTC Arrival Time: Aug 13, 2018 06:26:01.194530000 UTC
    Epoch Arrival Time: 1534141561.194530000
    [Time shift for this packet: 0.000000000 seconds]
    [Time delta from previous captured frame: 0.000000000 seconds]
    [Time delta from previous displayed frame: 0.000000000 seconds]
    [Time since reference or first frame: 0.000000000 seconds]
    Frame Number: 1
    Frame Length: 118 bytes (944 bits)
    Capture Length: 118 bytes (944 bits)
    [Frame is marked: False]
    [Frame is ignored: False]
    [Protocols in frame: eth:ethertype:ip:udp:skype]
    [Coloring Rule Name: UDP]
    [Coloring Rule String: udp]
  Ethernet II, Src: SunrichTechn_23:21:9b (00:0a:cd:23:21:9b), Dst: IPv4mcast_6f:ef:05 (01:00:5e:6f:ef:05
    Destination: IPv4mcast_6f:ef:05 (01:00:5e:6f:ef:05)
    Source: SunrichTechn_23:21:9b (00:0a:cd:23:21:9b)
    Type: IPv4 (0x0800)
  Internet Protocol Version 4, Src: 172.17.4.75, Dst: 239.239.239.5
    0100 .... = Version: 4
    .... 0101 = Header Length: 20 bytes (5)
    Differentiated Services Field: 0x00 (DSCP: CS0, ECN: Not-ECT)
    Total Length: 104
    Identification: 0x782b (30763)
    000. .... = Flags: 0x0
    ...0 0000 0000 0000 = Fragment Offset: 0
    Time to Live: 1
    Protocol: UDP (17)
    Header Checksum: 0xb208 [validation disabled]
    [Header checksum status: Unverified]
    Source Address: 172.17.4.75
    Destination Address: 239.239.239.5
  User Datagram Protocol, Src Port: 50431, Dst Port: 2423
    Source Port: 50431
    Destination Port: 2423
    Length: 84
    Checksum: 0x5b58 [unverified]
    [Checksum Status: Unverified]
    [Stream index: 0]
    [Timestamps]
    UDP payload (76 bytes)
  SKYPE
  [Community ID: 1:voj7nUaImcAbNR8EdO8CqBH4Ee8=]
```

Figure 3.18 – Example of ABB 800xA packet capture

- Honeywell Experion (Safety Manager):

```
∨ Frame 1: 68 bytes on wire (544 bits), 68 bytes captured (544 bits)
      Encapsulation type: Ethernet (1)
      Arrival Time: Dec  9, 2021 11:02:54.919935000 CET
      UTC Arrival Time: Dec  9, 2021 10:02:54.919935000 UTC
      Epoch Arrival Time: 1639044174.919935000
      [Time shift for this packet: 0.000000000 seconds]
      [Time delta from previous captured frame: 0.000000000 seconds]
      [Time delta from previous displayed frame: 0.000000000 seconds]
      [Time since reference or first frame: 0.000000000 seconds]
      Frame Number: 1
      Frame Length: 68 bytes (544 bits)
      Capture Length: 68 bytes (544 bits)
      [Frame is marked: False]
      [Frame is ignored: False]
      [Protocols in frame: eth:ethertype:ip:udp:data]
      [Coloring Rule Name: UDP]
      [Coloring Rule String: udp]
  ∨ Ethernet II, Src: SafetyManage_00:9e:ba (00:50:89:00:9e:ba), Dst: SafetyManage_00:9f:b3 (00:50:89:00:9f
    › Destination: SafetyManage_00:9f:b3 (00:50:89:00:9f:b3)
    › Source: SafetyManage_00:9e:ba (00:50:89:00:9e:ba)
      Type: IPv4 (0x0800)
  ∨ Internet Protocol Version 4, Src: 10.1.4.171, Dst: 10.1.4.173
      0100 .... = Version: 4
      .... 0101 = Header Length: 20 bytes (5)
    › Differentiated Services Field: 0x00 (DSCP: CS0, ECN: Not-ECT)
      Total Length: 54
      Identification: 0xadc0 (44480)
    › 000. .... = Flags: 0x0
      ...0 0000 0000 0000 = Fragment Offset: 0
      Time to Live: 64
      Protocol: UDP (17)
      Header Checksum: 0xaf9d [validation disabled]
      [Header checksum status: Unverified]
      Source Address: 10.1.4.171
      Destination Address: 10.1.4.173
  ∨ User Datagram Protocol, Src Port: 51020, Dst Port: 51030
      Source Port: 51020
      Destination Port: 51030
      Length: 34
    › Checksum: 0x0000 [zero-value ignored]
      [Stream index: 0]
    › [Timestamps]
      UDP payload (26 bytes)
  ∨ Data (26 bytes)
      Data: 010181000010bffb00009409282ec4000000000000000000000000
      [Length: 26]
      [Community ID: 1:m4eZJxc9RoOGTajw/tUAbWYw2FE=]
```

Figure 3.19 – Example of Honeywell Safety Manager packet capture

- Triconex (Schneider):

```
Frame 1: 48 bytes on wire (384 bits), 48 bytes captured (384 bits) on interface \Device\NPF_{98971CEC-2
    Section number: 1
    Interface id: 0 (\Device\NPF_{98971CEC-25CB-4C6C-ADFE-FC6003304D43})
    Encapsulation type: Ethernet (1)
    Arrival Time: Feb 22, 2018 15:29:18.818008000 CET
    UTC Arrival Time: Feb 22, 2018 14:29:18.818008000 UTC
    Epoch Arrival Time: 1519309758.818008000
    [Time shift for this packet: 0.000000000 seconds]
    [Time delta from previous captured frame: 0.000000000 seconds]
    [Time delta from previous displayed frame: 0.000000000 seconds]
    [Time since reference or first frame: 0.000000000 seconds]
    Frame Number: 1
    Frame Length: 48 bytes (384 bits)
    Capture Length: 48 bytes (384 bits)
    [Frame is marked: False]
    [Frame is ignored: False]
    [Protocols in frame: eth:ethertype:ip:udp:skype]
    [Coloring Rule Name: UDP]
    [Coloring Rule String: udp]
Ethernet II, Src: VMware_a3:9e:36 (00:0c:29:a3:9e:36), Dst: 40:00:00:00:00:02 (40:00:00:00:00:02)
    Destination: 40:00:00:00:00:02 (40:00:00:00:00:02)
    Source: VMware_a3:9e:36 (00:0c:29:a3:9e:36)
    Type: IPv4 (0x0800)
Internet Protocol Version 4, Src: 192.168.1.7, Dst: 192.168.1.2
    0100 .... = Version: 4
    .... 0101 = Header Length: 20 bytes (5)
    Differentiated Services Field: 0x00 (DSCP: CS0, ECN: Not-ECT)
    Total Length: 34
    Identification: 0x03d6 (982)
    000. .... = Flags: 0x0
    ...0 0000 0000 0000 = Fragment Offset: 0
    Time to Live: 128
    Protocol: UDP (17)
    Header Checksum: 0xb39b [validation disabled]
    [Header checksum status: Unverified]
    Source Address: 192.168.1.7
    Destination Address: 192.168.1.2
User Datagram Protocol, Src Port: 1078, Dst Port: 1502
    Source Port: 1078
    Destination Port: 1502
    Length: 14
    Checksum: 0x6bdc [unverified]
    [Checksum Status: Unverified]
    [Stream index: 0]
    [Timestamps]
    UDP payload (6 bytes)
SKYPE
    [Community ID: 1:Id/H4Sph/IrXFBNSfXaEHzHs8UU=]
```

Figure 3.20 – Example of Triconex packet capture

- Bently Nevada:

```
Frame 1: 76 bytes on wire (608 bits), 76 bytes captured (608 bits)
    Encapsulation type: Ethernet (1)
    Arrival Time: Mar 24, 2021 14:35:38.830962000 CET
    UTC Arrival Time: Mar 24, 2021 13:35:38.830962000 UTC
    Epoch Arrival Time: 1616592938.830962000
    [Time shift for this packet: 0.000000000 seconds]
    [Time delta from previous captured frame: 0.000000000 seconds]
    [Time delta from previous displayed frame: 0.000000000 seconds]
    [Time since reference or first frame: 0.000000000 seconds]
    Frame Number: 1
    Frame Length: 76 bytes (608 bits)
    Capture Length: 76 bytes (608 bits)
    [Frame is marked: False]
    [Frame is ignored: False]
    [Protocols in frame: eth:ethertype:ip:tcp]
    [Coloring Rule Name: TCP]
    [Coloring Rule String: tcp]
Ethernet II, Src: Dell_8b:21:8e (4c:d9:8f:8b:21:8e), Dst: Fortinet_09:0a:07 (00:09:0f:09:0a:07)
    Destination: Fortinet_09:0a:07 (00:09:0f:09:0a:07)
    Source: Dell_8b:21:8e (4c:d9:8f:8b:21:8e)
    Type: IPv4 (0x0800)
Internet Protocol Version 4, Src: 10.150.50.49, Dst: 149.182.206.176
    0100 .... = Version: 4
    .... 0101 = Header Length: 20 bytes (5)
    Differentiated Services Field: 0x00 (DSCP: CS0, ECN: Not-ECT)
    Total Length: 62
    Identification: 0x718a (29066)
    000. .... = Flags: 0x0
    ...0 0000 0000 0000 = Fragment Offset: 0
    Time to Live: 128
    Protocol: TCP (6)
    Header Checksum: 0x2802 [validation disabled]
    [Header checksum status: Unverified]
    Source Address: 10.150.50.49
    Destination Address: 149.182.206.176
Transmission Control Protocol, Src Port: 49968, Dst Port: 3500, Seq: 1, Ack: 1, Len: 22
    Source Port: 49968
    Destination Port: 3500
    [Stream index: 0]
    [Conversation completeness: Incomplete (12)]
    [TCP Segment Len: 22]
    Sequence Number: 1    (relative sequence number)
    Sequence Number (raw): 2485717204
    [Next Sequence Number: 23    (relative sequence number)]
    Acknowledgment Number: 1    (relative ack number)
    Acknowledgment number (raw): 3460917674
    0101 .... = Header Length: 20 bytes (5)
    Flags: 0x018 (PSH, ACK)
```

Figure 3.21 – Example of Bently Nevada packet capture

Based on multiple research findings and bulletins issued by the **Cybersecurity and Infrastructure Security Agency (CISA)**, the concept of *insecure by design* remains a significant challenge for many products and protocols currently in the market – with SIS standing out as a noteworthy example. There have been several recommendations and bulletins issued specifically aimed at improving the authentication, authorization, and encryption methods within these protocols, thus enhancing the security profile of SISs.

Lab exercise – TRITON

In this exercise, we will analyze the TRITON malware using Tricotools. The goal is to try to detect TRITON malware using specific indicators obtained during malware analysis performed in the Nozomi laboratory:

1. Download and install Wireshark from the appropriate website for your platform. For this exercise, we will use Tricotools, under `https://github.com/NozomiNetworks/tricotools`.

2. Run Wireshark and open the packet capture file you want to analyze.

3. On the menu bar, click on **Analyze** and select **Enabled Protocols...**.

4. Find the **Tristation** protocol dissector in the list and select it.

5. Click **Apply**, then **OK**.

6. Use the **Filters** field in the toolbar to sort the packets according to the `tristation.mid == 0x1f` filter to subset the packets that are identified as the TriStation protocol.

7. If you need to determine the source of the packets, use the `ip.src` or `ip.dst` filter to find the source IP address.

8. Refer to the link provided in *step 1* for a description of the individual packets. Try to investigate the packet stream for the following:

    *   The direction of communication

    *   Function codes translated as descriptive text.

    *   Extraction of transmitted PLC programs

    *   TRITON malware detection

    Additionally, you can analyze them one by one in the **Packet Details** window by analyzing the payload fields provided.

9. Once you are finished with your analysis, you can save your results as a text file or output them with the file export option in the **File** menu.

More detailed information about plugin installations can be found on the official web page:

`https://www.wireshark.org/docs/wsug_html_chunked/ChPluginFolders.html`

Summary

In this chapter, we addressed the need to properly secure an ICS by providing an overview of DiD layers, security design, and architecture, examining ICS components, and providing insight into real-life examples of SIS reference architectures (including Honeywell, Yokogawa, Emerson, ABB, Siemens, and Rockwell) and exploring the different safety network protocols that are available for secure communications.

The next chapter will focus on identifying attack vectors and entry points for attacking SISs, as well as techniques for thwarting such malicious activities.

Further reading

- Honeywell Experion: `https://www.arcweb.com/blog/honeywell-provides-more-flexible-scalable-safety-platform`

- Yokogawa: `https://instrumentationtools.com/yokogawa-dcs-and-sis/`

- Emerson: `https://www.emerson.com/documents/automation/white-paper-deltav-sis-cybersecurity-en-1262078.pdf`

- Rockwell: `https://www.chemanager-online.com/en/products/control-automation/plantpax-expansion-includes-integration-ics-triplex-sil-3-fault-tolerant`

- Siemens: `https://www.siemens.com/au/en/markets/oil-gas/sis.html`

- ABB: `https://iotsecuritynews.com/abb-system-800xa-base/`

- HIMA: `https://www.hima.com/en/products-services/sis-security-check`

- TRITON analysis: `https://www.mandiant.com/resources/blog/attackers-deploy-new-ics-attack-framework-triton`

- LOPA: `https://www.isa.org/intech-home/2016/january-february/features/safety-instrumented-system-sil-calculation`

- Purdue: `https://gca.isa.org/blog/excerpt-2-industrial-cybersecurity-case-studies-and-best-practices`

- ISA: `https://www.isa.org/standards-and-publications/isa-publications`

- NAMUR: `https://www.namur.net/en/`

- Wireshark: `https://www.wireshark.org/docs/wsug_html_chunked/ChPluginFolders.html`

- TriStation: `https://github.com/NozomiNetworks/tricotools`

- Awesome industrial protocols: `https://github.com/Orange-Cyberdefense/awesome-industrial-protocols`

- ICS protocols: `https://github.com/ITI/ICS-Security-Tools/blob/master/protocols/PORTS.md`

- List of MAC addresses for vendors with vendors' identities: `https://gist.github.com/aallan/b4bb86db86079509e6159810ae9bd3e4`

- Microsoft Defender for IoT – supported IoT, OT, ICS, and SCADA protocols: `https://learn.microsoft.com/en-us/azure/defender-for-iot/organizations/concept-supported-protocols`

Part 2:
Attacking and Securing SISs

Understanding effective SIS cybersecurity necessitates a grasp of threat actors' **Tactics, Techniques, and Procedures** (**TTPs**). *Chapter 4* offers a high-level overview of common attack methodologies and explores how process safety networks present a unique attack surface with various attack vectors.

A robust cybersecurity strategy hinges on proper design and a **Comprehensive Cybersecurity Management System** (**CSMS**). Deploying appropriate security controls is essential to enforce both logical and physical security. *Chapter 5* delves into practical security controls, emphasizing their importance in SIS deployment.

This part has the following chapters:

- *Chapter 4, Hacking Safety Instrumented Systems*
- *Chapter 5, Securing Safety Instrumented Systems*

This section provides the foundational knowledge and practical strategies necessary for securing SISs against sophisticated cyber threats.

4

Hacking Safety Instrumented Systems

The increasing complexity of modern Industrial Control Systems (ICSs) has opened up the doors to a number of new and emerging cybersecurity threats. As seen in the case of TRITON, BlackEnergy, Destroyer, and Stuxnet, to name but a few, advanced cyber attacks have the very real ability to cause major disruption and damage.

In response to these incidents, the process industry has been called upon to take greater preventive actions to ensure the ongoing security of these critical assets. However, too often, these initiatives have been temporary and compliance-oriented – as such, they have not fully addressed many of the underlying security issues.

Therefore, in order to protect mission critical systems such as ICSs and Safety Instrumented Systems (SISs), it is essential to understand potential threats and vulnerabilities associated with both. This chapter will examine various attack strategies employed by malicious actors, in the process revealing common attack channels in key areas. It also provides an overview of attack surfaces and a better understanding of the unique challenges faced by SISs in today's world.

Furthermore, this chapter discusses some security practices and countermeasures that can be taken to safeguard SISs from potential attacks. Ultimately, this chapter aims to demonstrate that, in light of increasing cyber threat levels, a proactive approach *must* be taken by the wider industry to ensure safe and reliable operations.

We will be covering the following topics in this chapter:

- ICS attacks
- Understanding the SIS attack surface
- Attacking the SIS controller
- (P)0wning the **Safety Engineering Workstation (S-EWS)**

- Abusing the Instrument Asset Management System (**IAMS**)
- Replaying traffic
- Reverse engineering of a transmitter of field devices
- Bypassing a key switch
- Putting it all together
- Lab exercise – ReeR MOSAIC M1S safety PLC security assessment

ICS attacks

Traditional ICSs were developed with the aim of providing dependability, durability, economic efficiency, and safe operations. The increasing implementation and digitalization of control systems, coupled with integration between vertical and horizontal systems – such as IT, telecom, and the cloud – has enabled the establishment of cyber interdependencies, which have become a source of unintended exposures and weaknesses, often difficult to anticipate.

The introduction of a number of new technologies has improved efficiencies, yet at the same time also raised concerns regarding the vulnerability of plants to both random cyber failures and security infringements. As a result, physical processes and equipment have become open to exploitation. Understanding the implications of this class of cyber attack, in terms of an attacker's ability to cause physical damage, is invaluable when it comes to grasping how to execute a successful cyber-physical attack.

The following diagram conveys the distinct components of a cyber-physical attack:

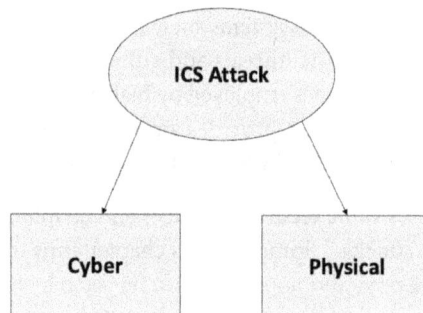

Figure 4.1 – Cyber-physical attack

The primary objective of a cyber attack is to infiltrate infrastructures or parts of them. However, a physical (or engineering) attack is designed to disrupt, alter, or demolish the operational process and its equipment. Consequently, it has notable physical implications.

Process (physical) attacks

As discussed previously, process control involves employing methods for controlling the variables of the process or processes used in the production of a certain product. Through this intervention, parameters such as process inputs, process parameters, and process outcomes can be modified. The ultimate intention is to minimize the variability of the process, thereby improving product quality, production rates, process efficiency, environmental protection, and the safety of personnel and equipment.

Engineering attacks within the process industry typically target key process functions that both manage the plant and ensure process control and safety are maintained in tandem. These functions could be used in cyber-physical attacks in several ways.

The attacker must possess relevant knowledge in terms of signal processing, control principles, physics, mechanics, and failure conditions, to name just a few. In addition, the failure modes of the various ICS types differ greatly and must all be understood in order to gain the necessary level of insight to safeguard against attack.

A well-crafted process attack will require good engineering knowledge as well as hands-on experience with process control and safety. This includes, but is not limited to, the following:

- Understanding the process: It is important to fully understand process control and safety and its workings, as well as the results it produces.

- Pinpointing operating parameters: Understanding specific conditions essential for controlling the process. The operating criteria for pressure, temperature, and flow rate should all be known and always carefully controlled. Additional parameters that are commonly measured include pH, humidity, level, concentration, viscosity, conductivity, turbidity, redox potential, electrical behavior, and flammability.

- Identifying hazardous conditions: Investigating the risk area and the variables that can cause safety issues.

- Locating measurement points: Identifying measurable parameters and where they are monitored to understand how the system is regulated.

- Identifying control method: Understanding how operating parameters are managed. On/off is one control method, while continuous regulation entails operating with **proportional (P)**, **integral (I)**, and **derivative (D)** methods or any combination of the three.

- Control system technology: Knowledge of **Distributed Control Systems (DCSs)**/SISs helps, as these systems have static architecture and engineering practices: vendor and **Original Equipment Manufacturer (OEM)** documentation is widely accessible.

- Control logic programming: Developing these skills helps, especially when it comes to modification to control or safety criteria of the system.

Malicious users can disrupt processes in many ways if they have access to the ICS network; they may not require hacking artifacts for this purpose as the majority of these actions can be carried out using native control systems' toolboxes and utilities that are shipped as part of installation.

There are numerous ways that attack vectors could be initiated as part of an ICS attack, as depicted in the following diagram:

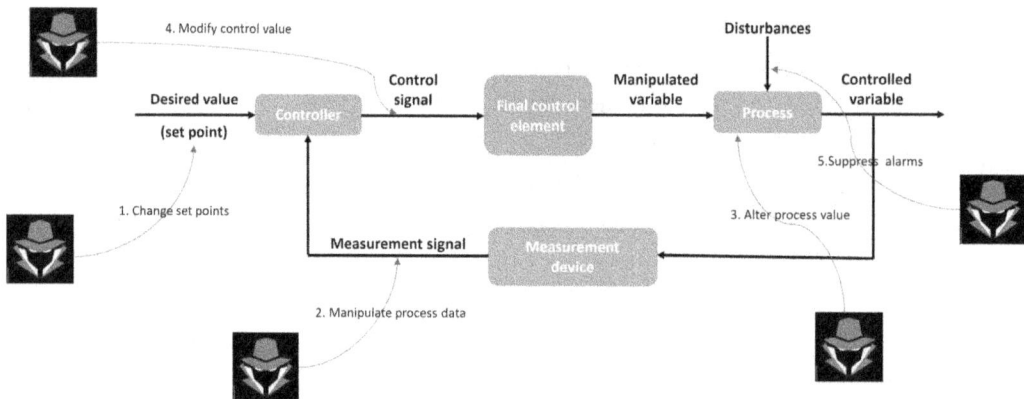

Figure 4.2 – Process attack vectors

Many stealthier, successful attacks will focus on at least one of the five areas listed or will combine some of these actions using solely engineering tools and utilities that are deemed to be legitimate:

- Configuration parameters: This allows control parameters such as set points to be overridden, creating new points that can affect the stability of the processing program

- Manipulate process data: This can give attackers access to confidential information and allow them to alter operations that may, in turn, cause process upset or top events

- Alter process values: This can lead to inaccurate measurements and unpredictable results

- Modify control values: This can disrupt the correct functioning of the system, as these values determine the control performance that is given to the user

- Suppress alarms: This can allow malicious users to exploit the system unnoticed, as they don't have to respond to any alarm triggers

Some examples: the increased pressure in tanks or pipelines can lead to an explosion, which is a threat to both people and assets; similarly, the temperature can be manipulated to cause explosions in extreme cases. Humidity, tank gauging, and level readings can all be manipulated to affect the environment—for instance, some tanks contain chemicals that are dangerous at certain levels, and the manipulation of the readings could lead to a hazardous situation.

Ultimately, these process control and safety functions are integral to the overall safety of critical infrastructure and machinery. If they are manipulated and used as tools in cyber-physical attacks, the results can be catastrophic. Fortunately, all known cyber-physical incidents focused on sabotage or degradation of a process instead of causing hazards.

Next are some examples of cyber-physical attack scenarios that are used as reference points in many safety-case exercises and cyber **Control Systems Hazard and Operability Study (CHAZOP)** studies. Unlike IT systems, the impact will be on physical processes, as depicted here:

- **Boiler blowout**: This cyber-physical attack involves the attacker using a weaponized exploit to plan and execute a damaging boiler explosion scenario. The scenario starts with the attacker stopping the feedwater into the boiler, causing the drum to overheat. The attacker then reintroduces the feedwater back into the boiler, causing an explosion. This explosion leads to immense property damage, a risk to life, and a threat to connected systems:

Figure 4.3 – Boiler installation

- **Ammonia tank explosion**: A cyber-physical attack might lead to an explosion at an ammonia plant by manipulating the heating process – loss of cooling leading to a rise in pressure that might result in an explosion, disabling alarms and safety systems, and increasing concentrations of carbon monoxide in the methanator in combination with an ignition source in the vicinity, leading to intense destruction and potential loss of life as ammonia extracts hydrogen from the air, making it difficult to breathe:

Figure 4.4 – Ammonia tank

- **Turbine overspeed:** The attacker might successfully carry out actions to disable turbine overspeed shutdowns and disconnect turbine loads, thus allowing the turbine to speed up rapidly and reach dangerous operation levels. The consequences of this attack could include a devastating breakdown of the turbine system and an inability to supply power to a site or a region:

Figure 4.5 – Gas turbine

Of course, these three examples have one element in common: SISs. SISs will likely be deployed to safeguard the process and could also be an interesting target of choice depending on the attacker's profile and motivation – whether that be economic, political, and/or military interests.

In recent years, attackers have been inclined to shift their operations down toward control and safety zones (level 3 to level 1). This phenomenon, known as the race-to-the-bottom trend, is especially visible between threat actors and defenders of ICSs. As owners of assets increase their investments in hardening operating systems and networks and deploying ICS anomaly detection and monitoring, attackers are still managing to find their pathway into the "crown jewels" and penetrate the control equipment.

The following diagram shows how attacks have been executed to attack process control facilities:

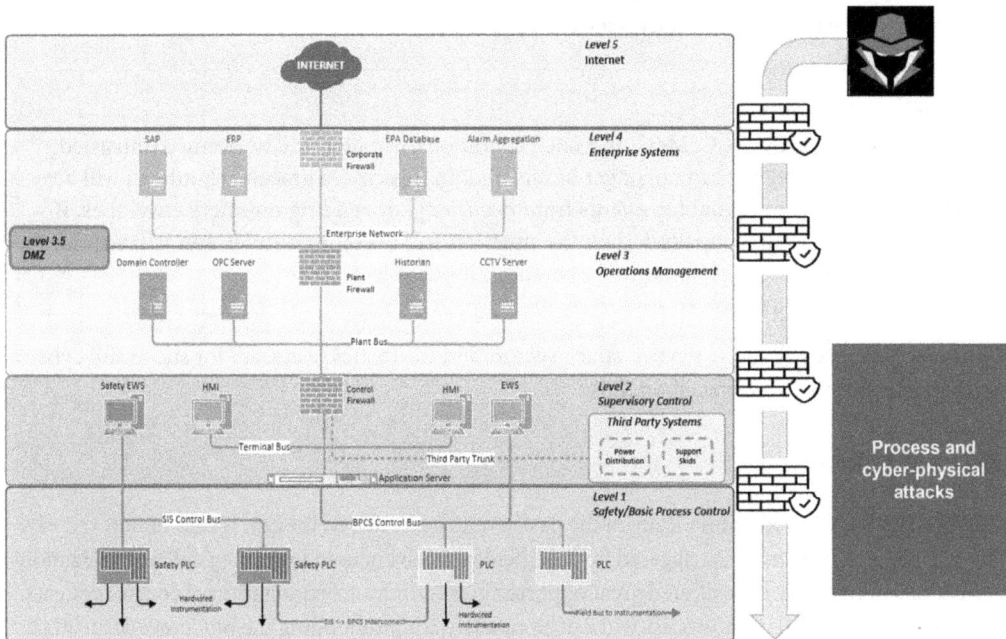

Figure 4.6 – Race-to-the-bottom attack trends

In this type of attack, various artifacts will be applied to manipulate or take control of process functions. Some of these attacks might go undetected depending on the stealth abilities of the attacker. This can be achieved through the following tactics:

- **Loss of View (LoV):** The attacker is potentially able to prevent legitimate users from monitoring the status of the system, assets, or processes

- **Manipulation of View (MoV):** The attacker is potentially able to modify the output from legitimate monitoring software or nodes in order to mislead users or create confusion around the system's status, assets, or processes

- Denial of Control (DoC): The attacker is potentially able to prevent legitimate access to the regulation and operation of the system, assets, or processes

- **Manipulation of Control (MoC)**: The attacker is potentially able to modify the input of legitimate control software or nodes in order to change the system's operations

- **Loss of Control (LoC)**: The attacker is potentially able to take away legitimate users' access to the regulation and operation of the system, assets, or processes

Key components responsible for monitoring, controlling, and safeguarding are necessary for the actions outlined previously. As covered in Chapter 3, these components (**Human-Machine Interface (HMI)**, S-EWS, DCS/SIS controllers, IAMS) – often referred to as "crown jewels" – are likely to be the primary target or part of a bigger attack scheme.

> **Important note**
>
> Under attack (LoV, MoV, DoC, MoC, and LoC) conditions, a system will be deemed untrusted. Therefore, process integrity can no longer be ensured. In this case, a graceful shutdown will be initiated to avert any potential top events from occurring, depending on safety envelopes. If the safety function is also impaired, then the situation will become complex and will require remediation and scaling back to normal operations if possible.

In the next section, we will discuss cyber-attack vectors and the tactics necessary for successful cyber-physical attacks, with a focus placed on SISs.

Cyber attacks

The ICS domain has evolved dramatically in the past several years due to the emergence of new security threats, catalyzing global worries triggered by past incidents: advances in technological modernization and extended connectivity have offered great opportunities for remote exploitation of control systems. Nonetheless, efforts to disrupt a process without completely comprehending the outcomes of an attack on a physical process are likely to lead to a minor disturbance instead of any true damage.

Consequently, to achieve a desired cyber-physical effect, an attacker needs to possess a strong understanding, capabilities, and resources, distinct from what is usually utilized within the IT field: for instance, committing a tailored attack to bring the process into a vulnerable state or waiting for a process to reach a certain state prior to executing an attack is necessary.

In the past, ICS infiltration was perceived as a specialized talent usually owned by state-backed malicious actors; SIS security knowledge was exclusive to a select group. This status quo has been questioned over the past decade thanks to the greater availability of ICS equipment, open-source attack tools, and information, all of which permit less accomplished cybercriminals to engage in the ICS realm.

In the next few paragraphs, we will learn more about the methodology and attack vectors that are being used to target SISs.

Attackers now use a variety of techniques to gain access to ICSs, ranging from social engineering to physical attacks. As such, it is important for organizations to understand the varying methods that attackers can utilize and to be aware of steps that can be taken to better protect their ICS.

The Cyber Kill Chain is an analytical framework that explains the process cybercriminals use to target and attack critical infrastructure. Through this chain, it is possible to analyze and anticipate potential intrusions, as criminals follow the same techniques to successfully penetrate systems.

The Cyber Kill Chain is a multi-phase model used to describe the stages of an attack, from the initial compromise and execution of malicious activity all the way to the attackers' objectives being achieved. The following diagram depicts the Cyber Kill Chain phases:

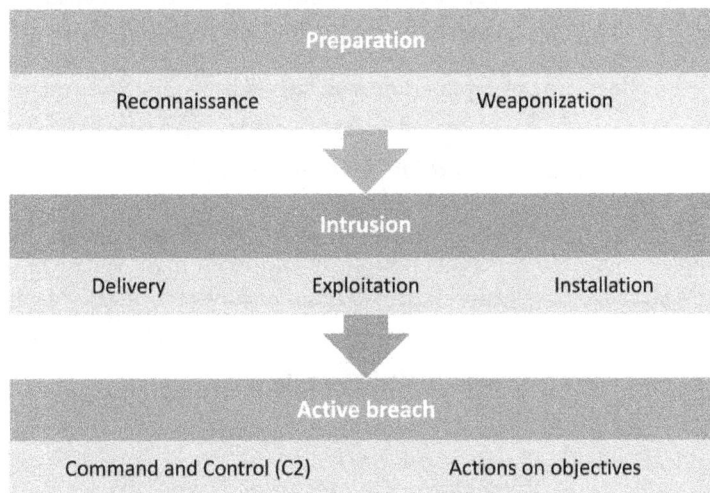

Figure 4.7 – Cyber Kill Chain

The Cyber Kill Chain is divided into seven stages that describe the process and provide insight into the tactics and methods used by malicious actors to carry out their attacks. As we dive deeper into each stage, we gain a better understanding of the steps involved in the Cyber Kill Chain, allowing for better preparation and defense against potential threats:

- **Reconnaissance**: The first step of the chain, reconnaissance, is when malicious actors scan for security vulnerabilities and gather information through public information or deploying social engineering techniques.

- **Weaponization**: The second step, weaponization, is when attackers use the knowledge and insight gained from the previous step to create malicious software that can be used use against the target, including malware or zero-day exploits.

- **Delivery**: Delivery is when malware is sent to the victim. It's usually through an email attachment, a link to a malicious website, or a file downloaded from a malicious website or from USB sticks within ICS environments.

- **Exploitation**: Exploitation is the stage when attackers attempt to gain access to the system by exploiting the vulnerability they have identified.

- **Installation**: Installation is when the attackers install malicious software on the targeted system.

- **Command and Control (C2)**: The next step, C2, is when attackers use this malicious software to control the target system from their own computers, and data or trade secrets can be stolen.

- **Actions on objective**: The final step, actions on objective, is when an attacker finally reaches their objective by either collecting sensitive information or disrupting the target system.

Now we have the methodology, we can explore common targets, especially SISs and their interfaces.

Understanding the SIS attack surface

To make an SIS more secure, we need to clearly understand the attack surface – this includes all entry points where an unauthorized user could potentially gain access to the system or interfaces and exploit vulnerabilities for malicious purposes. An SIS could be subject to attacks via direct network connections, remote access software, portable media, or even supply chain threats, among others.

First, let's explore SIS interfaces and **Systems Under Consideration (SuCs)**.

SuCs

When evaluating ICSs, especially SISs, it is vital to understand the scope and any dependencies as well as system boundaries. This requires the examination of all interfaces and entry points, including physical and digital venues.

As per the *IEC 62443-3-2* definition, the SuC encompasses anything from **Basic Process Control Systems (BPCSs)**, **Distributed Control Systems (DCSs)**, SISs, and **Supervisory Control And Data Acquisition (SCADA)**, ICS, or **Industrial Automation and Control System (IACS)** product supplier packages. This may even include emerging technologies such as the **Industrial Internet of Things (IIoT)** or cloud-based solutions. The SuC described with the Purdue Enterprise Reference Architecture model spans from Level 0 up to Level 3.5 **Demilitarized Zone** (DMZ). To define the SuC, an organization can develop system inventory, architecture diagrams, network diagrams, and data flows to determine the ICS/IACS assets involved.

Adequately describing the SuC is of paramount importance when an organization begins to plan for an ICS cybersecurity risk assessment. It provides the baseline and the scope for identifying and understanding associated risks for the respective system, as well as the ability to effectively implement appropriate security controls to protect it. Next is an example of a SuC for BPCSs and SISs:

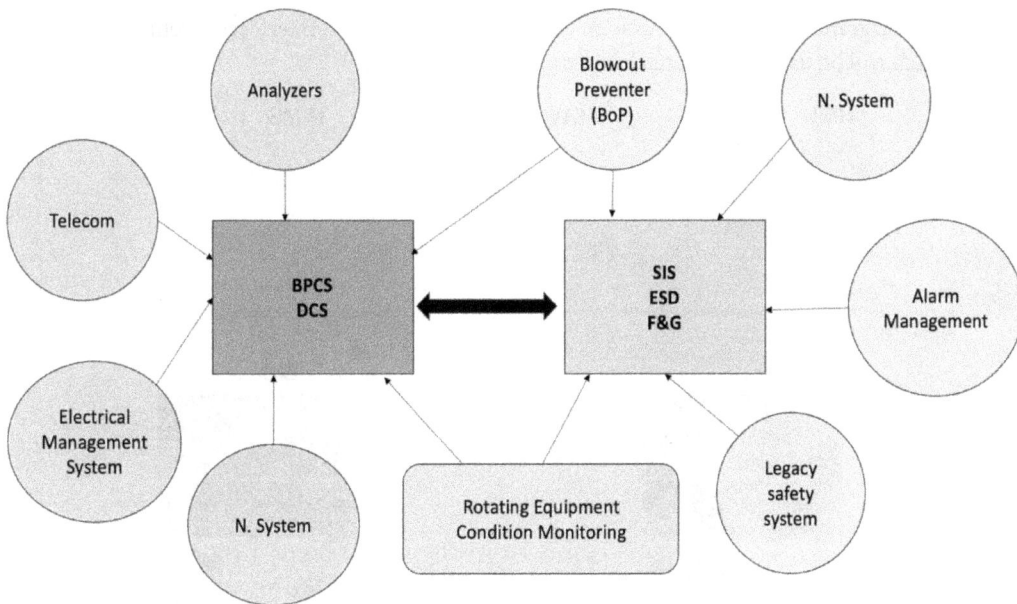

Figure 4.8 – SuC example for BPCSs and SISs

The adoption of a SuC is still evolving and faces a number of challenges, including the following:

- Long-established elements of legacy systems throughout assets.

- The absence of a clear Purdue Enterprise Reference Architecture model and system boundaries.

- The complexity of IT and OT components that existed in decades past, and a lack of clear ownership (pre-IT/OT convergence era).

- The rapid expansion of automation systems and increased ties to external borders, as well as an exponential rise in open industrial protocols, the introduction of new smart devices/IoT/ IIoT sensors, and the emergence of remote operation and maintenance activities. All of these have raised interdependence among various systems.

- Little to no Management of Change (MOC) procedures and **Document Management Systems (DMSs)** for decades-old systems.

- The introduction of disruptive automation protocols, next-generation technologies, and the accompanying features and design of legacy systems.

The following diagram shows a typical system with its components and interfaces – defined as a SuC for SIS – which will be used for the attacks described in this chapter:

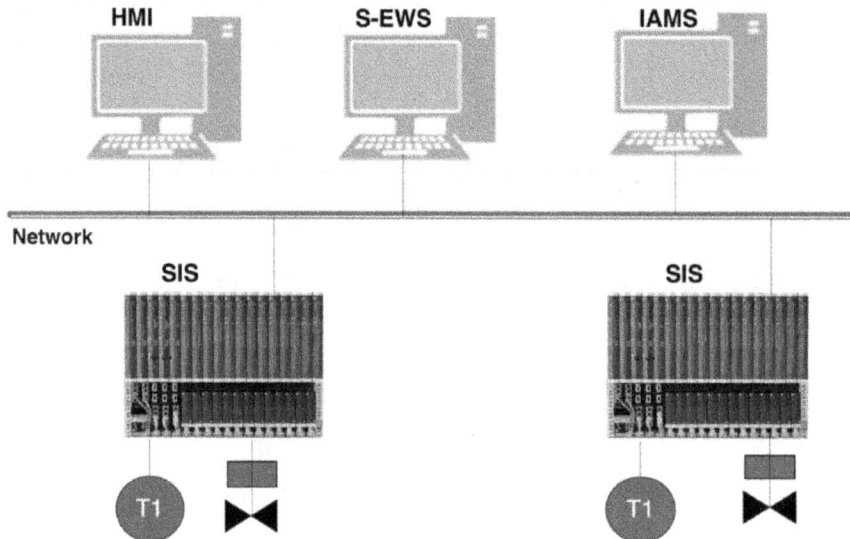

Figure 4.9 – SuC

The next section explains how these systems can be attacked based on past ICS incidents.

Abusing the HMI

Gaining unauthorized command and control of an ICS can be achieved by employing the features of an HMI console. This may be an embedded HMI within a control zone or a centralized C2 system such as a DCS, SCADA system, or SIS.

An easy way to manipulate the controls is through its console interface. Rather than attacking via the industrial network, an attacker can exploit a known device vulnerability in order to install remote access to the console, which can enable the hacker to completely take control of it. Penetration tools and frameworks can be used to exploit the target system and utilized to insert payloads of choice. This will give an attacker control over whatever the console is used to manage, regardless of knowledge of industrial protocols and ladder logic of control system operations.

The following diagram depicts example entry points for both HMI attacks and potentially also pivoting to other systems in the same network or bus:

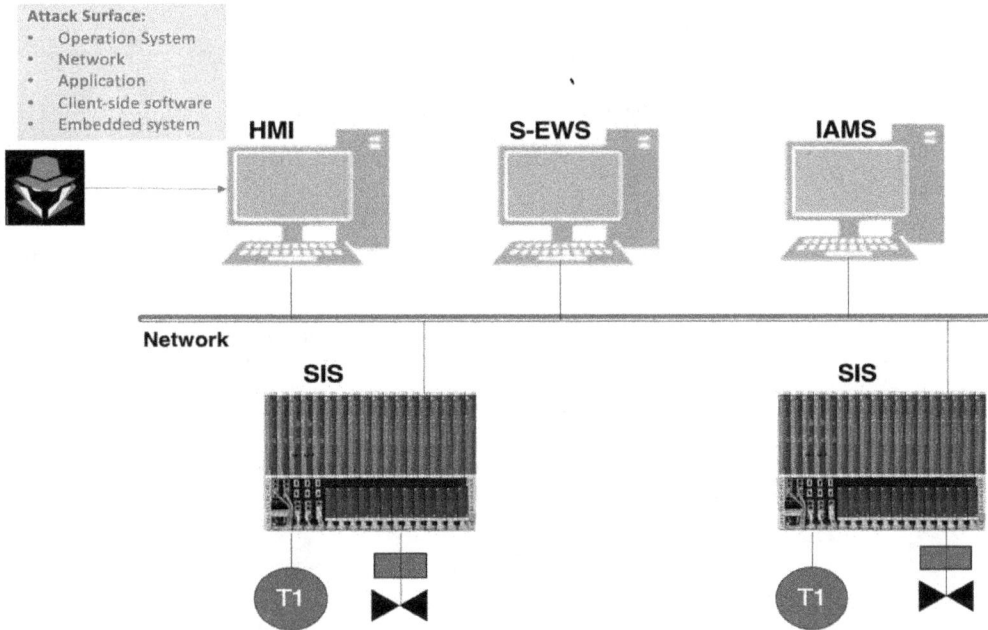

Figure 4.10 – HMI attack surface

HMIs have a wide attack surface that can be used to compromise the system and potentially jump to other hosts. The following table presents some possible attack vectors:

Attack vector	Attack path
Operating system (OS)	• Exploiting vulnerable patches in the OS • Using Trojan Horses and malicious programs designed to exploit the OS • Manipulating the OS configurations to gain unauthorized access • Using rootkits and remote administration tools • Leveraging **Living Off The Land (LOTL)**
Network	• **Distributed Denial-of-Service (DDoS)** attacks • Spoofing, manufacturing malicious packets, or manipulation of data • Unauthorized access across multiple networks • Exploiting existing network routing protocols

Attack vector	Attack path
Application	• Conditions in the application can be exploited to access data stored on the system • Privilege escalation using application flaws • Identification and authentication attacks • Malicious inputs that could cause unexpected behavior in the application • Conjunction attack to exploit authentication, authorization, and perimeter vulnerabilities
Client-side software	• Downloading malicious objects from malicious websites (although internet access should be prohibited for an HMI) • Unauthorized access to client-side software • Manipulating input validation processes and forms to access confidential information
Embedded systems and hardware	• Manipulation of software and hardware components • Exploiting known hardware vulnerabilities • Manipulating communications between components • Reverse engineering and hardware tampering (including physical access)
Third-party components	• Inclusion of malicious third-party components • Exploiting vulnerabilities in third-party components • SQL injection attacks on third-party databases • Untraditional attacks such as **Man-in-the-Middle** (**MITM**) attacks

Table 4.1 – Examples of HMI attack vectors

Once the HMI is compromised, the graphical user interface does not require a lot of expertise to navigate – click buttons and change values on a console – as this is designed for easy use.

Once the HMI is compromised, the attacker will focus on controlling the "crown jewels" by means of gaining access to the controller.

> **Important note**
> LOTL refers to an attack technique characterized by the use of fileless malware or **LOTL binaries** (**LOLBins**), wherein the cybercriminal leverages native, legitimate tools already present in the victim's system to perpetuate and escalate the attack. For ICSs, there are many tools and utilities that can be leveraged for the same purpose.

The next section covers multiple attack surfaces and relevant attack scenarios designed to compromise SIS controllers.

Attacking the SIS controller

Attackers can gain access to the controller through a variety of methods, including brute-force password cracking, dictionary attacks, buffer overflow attacks, and serial port sniffing. Attackers can also compromise the controller OS by utilizing malicious code such as malware. This malicious code is designed to penetrate the system and transmit sensitive data to attackers for further exploitation.

Once malicious code is installed, an attacker can use the compromised system to launch attacks on other connected systems.

SIS controllers communicate directly with the HMI, EWS, and IAMS. These interfaces can be interesting entry points to attack SIS controllers:

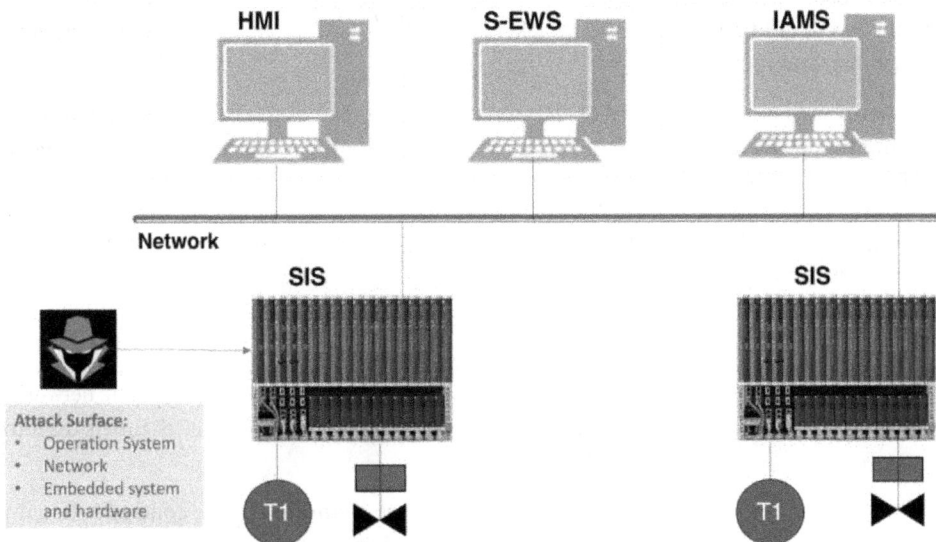

Figure 4.11 – Safety controller attack surface

In this context, the safety controller attack surface has four distinct layers:

- The logic layer operates at the uppermost degree of abstraction within a **Programmable Logic Controller** (**PLC**) system to dictate the unique functions of devices. This layer carries out a logic-based program cyclically through the use of a runtime system. Typically, the logic program is developed by coders at an EWS before being transferred to the PLC. This program can be written in a variety of coding languages that adhere to the *IEC 61131-3* protocol, including but not limited to ladder logic, the **Structured Text** (**ST**) language, or a **Function Block Diagram** (**FBD**).

- The task layer represents the highest level of firmware abstraction, and it comprises tasks assigned to facilitating communication, performing logic programs, and debugging. The task layer's crucial component is the runtime system, which handles a range of functions, including communicating with management systems, operating logic programs, and managing physical **Inputs and Outputs** (**I/Os**). Furthermore, specific PLC devices may also support other tasks such as web services, **File Transfer Protocol** (**FTP**), or Telnet. As the primary portal for external engagement, this layer is particularly vulnerable to cyber threats.

- The kernel layer, which forms an essential part of the firmware, is a significant abstract layer that is often found in devices powered by general-purpose operating systems. However, its presence may not be as distinct in some other systems. This layer's duties include governing peripheral activities, scheduling and backing up tasks, along with executing lower-level functions such as the initialization of the bootloader and loading of the operating system. Potential vulnerabilities in this layer could be exploited by attackers, enabling them to obtain higher privileges and launch harmful rootkits.

- The hardware layer is composed of electronic elements such as microprocessors, temporary memory, and permanent storage, all housed in microchips. These components play an instrumental role in ensuring the smooth operation of the PLC.

The following table provides a high-level breakdown of attack vectors and possible pathways for the execution of attacks on controllers:

Attack vector	Attack path
Network	<ul><li>Gain unauthorized access to the controller either by using public networks or by exploiting protocol flaws in the communication channels used to access the controller.</li><li>Use external services or interfaces connected to the controller to gain unauthorized access to the controller.</li></ul>

Attack vector	Attack path
Operating system	• Poor authentication and authorization schemes can make it easier for attackers to gain unauthorized access to the PLC controller. • Old or vulnerable firmware versions can provide attackers with an entry point into the controller or allow them to make malicious modifications to the controller's operating environment. • Access to the controller by exploiting a user's credentials such as passwords, keys, or web interface. • Unpatched operating system vulnerabilities can provide an opening for attackers to gain access to the controller.
Embedded systems and hardware	• Embedded programming to alter firmware settings that control how the controller interacts with other devices and DCSs/IAMSs. Exploitation of embedded systems can also lead attackers to take control of command functions on the controller itself. • Direct physical access to the controller can allow attackers to access backdoor elements of the system, install malicious code and hardware, or gain access and weaponize the controller.

Table 4.2 – SIS controller attack vector and path

Now we have explored the possible paths to exploit an SIS controller, we will dig deeper to better understand how this relates to other systems as defined in the SuC.

(P)0wning the S-EWS

Vectors for compromising an EWS may be similar to those used with earlier HMI systems. This is due to the consistent system management that exists across hosts. The significant factor that needs to be considered is the relative value of assets on the EWS compared to assets on the HMI.

The HMI is a bidirectional read/write instrument for the process being controlled; however, many systems presently involve **Role-Based Access Control** (**RBAC**), which may be limited if multiple operators and many plant units are involved.

The EWS and S-EWS generally include the ability to configure RBAC. They also provide the necessary tools to directly access, adjust, and improve the primary control equipment (PLC, BPCS, SIS, **Intelligent Electronic Device** (**IED**), and so on). Moreover, the EWS contains important confidential documents related to the ICS design, configuration, and operation, which makes it more valuable than the typical HMI. *Figure 4.12* illustrates various attack vectors and relevant scenarios for exploiting an S-EWS:

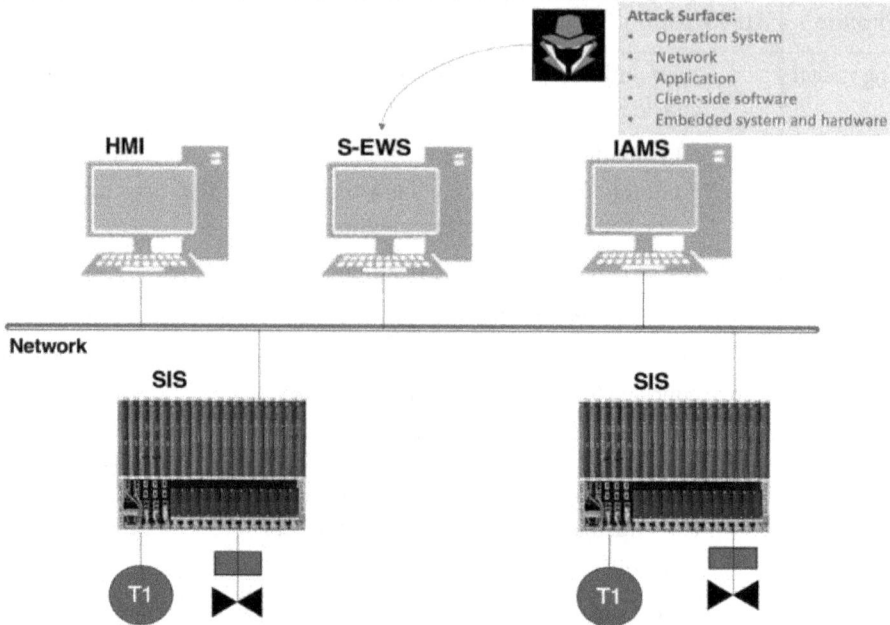

Figure 4.12 – S-EWS attack surface

Since the S-EWS is running on Windows machines and is equipped with software to program logic solvers and conduct changes of setpoints, this makes the S-EWS an attractive target – as we have learned from past incidents.

The S-EWS communicates with various systems and has similar weaknesses; as such, the attack surface can be exploited in the same way as the IAMS or HMI, which will be covered next.

Abusing the IAMS

IAMSs are software systems used to monitor and manage a wide range of industrial equipment and systems, such as SISs. IAMSs are critical in ensuring both the safety and availability of industrial systems. Yet, due to their importance, malicious users can utilize them as a means to carry out cyber attacks against industrial safety and control systems.

Malicious users can deploy various means to carry out cyber attacks against IAMSs. These include a variety of malicious software such as viruses and Trojans, as well as social engineering attacks. Using malicious software, malicious users can manipulate IAMSs to steal data, delete or corrupt data, install malicious software, disable legitimate software, or compromise network access. They can also use malicious software to gain access to SISs, which can lead to dangerous results.

In the case of social engineering attacks, malicious users can use techniques such as phishing and spear phishing to lure unsuspecting users to provide confidential information or access to systems. Furthermore, malicious users also target networked IAMSs in order to gain access to SISs.

By using so-called MITM attacks, they can quickly intercept data that flows between system components within a network. This enables malicious users to gain access to vital system functions, such as SISs, and to control them without the knowledge of others.

A successful cyber attack on IAMSs also enables malicious users to make unauthorized changes to networks or systems. For example, a malicious user may be able to tamper with IAMS configuration settings such as software versions, system input and output parameters, or personnel permissions. These changes can create dangerous circumstances and possibly lead to a critical malfunction of SISs.

In addition, malicious users can use IAMSs as a means to gain access to other systems that may be connected to the same network. A malicious user may be able to exploit an IAMS to gain access to other system components such as HMIs, **Process Control Systems (PCSs)**, and even databases: this could lead to significant data loss or unauthorized changes to system parameters. *Figure 4.13* depicts the attack surface along with relevant scenarios for compromising IAMSs:

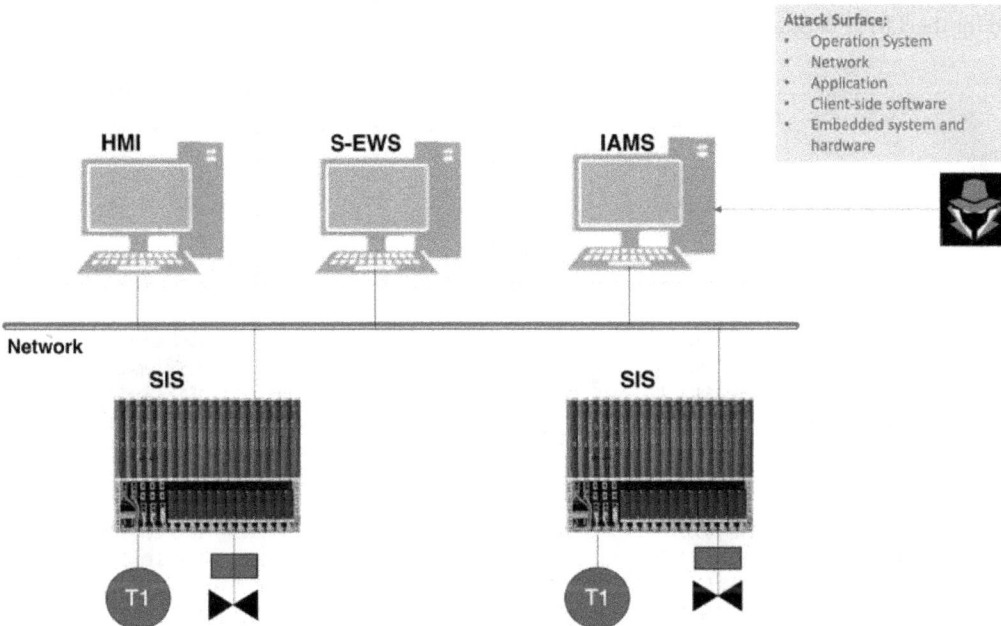

Figure 4.13 – IAMS attack surface

> **Important note**
> The attack vectors for HMIs and S-EWSs are the same as for IAMSs since they have the same interfaces.

Next, we will focus on packet capture for relevant safety protocols and demonstrate how replay attacks can be carried out.

Replaying traffic

A replay attack is a form of cybersecurity threat where a malicious actor intercepts and records communication data, only to retransmit it later. The system or network under attack assumes the retransmitted or "replayed" data to be legitimate. In the context of an SIS, an attacker could use this method to manipulate and inject harmful data streams. Essentially, they "replay" commands in an attempt to trick the system into executing unintended actions, potentially leading to system disruptions or even safety incidents if not properly mitigated.

Replaying network traffic requires the capture of real-time data from a network protocol using tools such as Wireshark and tcpdump. Replay attacks involve the modification of the traffic to mimic a malicious client, allowing for the injection of malicious or unauthorized requests.

An extract for Triconex **User Data Protocol (UDP)** packet capture (**1502/udp**) is shown in the following screenshot. This shows how the I/O and tag values could be changed and manipulated as used for the TRITON attacks:

No.	Time	Source	Destination	Protocol	Length	Info
24	2.234188	192.168.1.2	192.168.1.7	UDP	120	1502 → 1155 Len=78
25	2.234492	192.168.1.7	192.168.1.2	UDP	58	1155 → 1502 Len=16
26	2.415873	192.168.1.2	192.168.1.7	UDP	244	1502 → 1155 Len=202
27	2.459861	192.168.1.7	192.168.1.2	UDP	100	1155 → 1502 Len=58
28	2.630156	192.168.1.2	192.168.1.7	UDP	120	1502 → 1155 Len=78
29	2.630513	192.168.1.7	192.168.1.2	UDP	58	1155 → 1502 Len=16
30	2.811978	192.168.1.2	192.168.1.7	UDP	244	1502 → 1155 Len=202
31	2.864083	192.168.1.7	192.168.1.2	UDP	100	1155 → 1502 Len=58
32	3.008209	192.168.1.2	192.168.1.7	UDP	120	1502 → 1155 Len=78
33	3.008767	192.168.1.7	192.168.1.2	UDP	58	1155 → 1502 Len=16
34	3.208073	192.168.1.2	192.168.1.7	UDP	244	1502 → 1155 Len=202
35	3.279255	192.168.1.7	192.168.1.2	UDP	100	1155 → 1502 Len=58
36	3.412359	192.168.1.2	192.168.1.7	UDP	120	1502 → 1155 Len=78
37	3.412598	192.168.1.7	192.168.1.2	UDP	58	1155 → 1502 Len=16
38	3.630145	192.168.1.2	192.168.1.7	UDP	244	1502 → 1155 Len=202
39	3.686472	192.168.1.7	192.168.1.2	UDP	100	1155 → 1502 Len=58
40	3.808312	192.168.1.2	192.168.1.7	UDP	120	1502 → 1155 Len=78
41	3.808540	192.168.1.7	192.168.1.2	UDP	58	1155 → 1502 Len=16
42	4.008355	192.168.1.2	192.168.1.7	UDP	244	1502 → 1155 Len=202
43	4.084936	192.168.1.7	192.168.1.2	UDP	100	1155 → 1502 Len=58
44	4.214078	192.168.1.2	192.168.1.7	UDP	120	1502 → 1155 Len=78
45	4.214386	192.168.1.7	192.168.1.2	UDP	58	1155 → 1502 Len=16
46	4.414173	192.168.1.2	192.168.1.7	UDP	244	1502 → 1155 Len=202

Figure 4.14 – Triconex 1502/UDP traffic

A screenshot of data (tag changes) from Triconex packet capture is shown as follows:

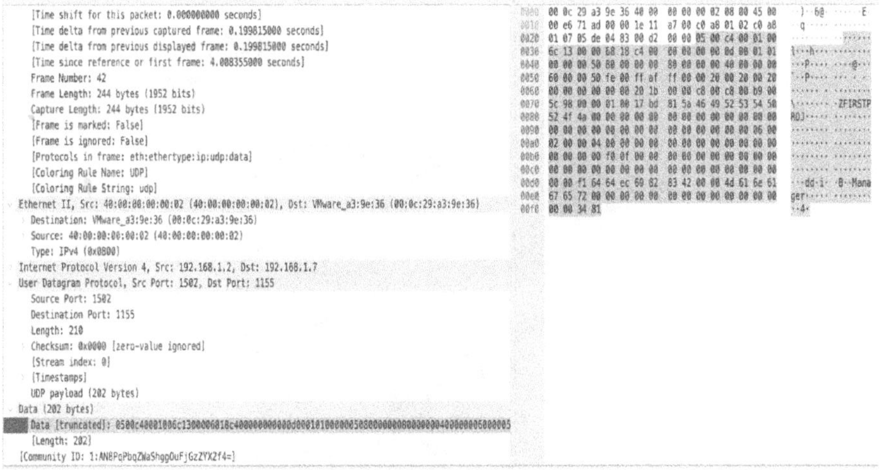

Figure 4.15 – Triconex change I/O and tag values

As demonstrated in this packet capture for Triconex traffic on port 1502/udp, it is clear that the majority of ICS protocols do not provide any security mechanism to prevent replay traffic as the protocol stack does not provide authentication, authorization, or encryption. Therefore, it's relatively easy to intercept traffic and manipulate certain values without getting noticed. The next section explains the process of reverse engineering a transmitter.

Reverse engineering a transmitter of field devices

Firmware is the backbone of most electronic components and devices. It is also known as embedded software and is responsible for controlling the behavior of devices. Firmware is used to define how the device should operate under various conditions, such as system operations, temperature, and environment. The first step to reverse engineer firmware is to create a copy of the existing firmware.

To do this, software is used to perform a *read* operation of the firmware, which allows a bit-for-bit memory image to be created. This memory image is then saved to a file for later analysis. Once the memory image has been created, it is analyzed by an automated tool using static binary analysis. This analysis will extract the program code, library functions, and other elements from the firmware. This provides a baseline for understanding the behavior of the firmware code as it is written. The next step is to perform a dynamic binary analysis. Here, the original firmware is run within a virtual environment and monitored, providing more in-depth insights. This allows for the identification of functions and variables, as well as any hidden or obfuscated code.

Finally, the results of the analysis can then be used to create custom firmware that is tailor-made for the specific needs of the device. This is often achieved using external development tools such as compilers or assemblers to create the custom firmware.

The following photo shows the hardware connection to the JTAG interface of a transmitter to extract and dump firmware for analysis:

Figure 4.16 – Transmitter firmware extraction

As depicted in the following screenshot, connecting to the transmitter using AVR Dragon confirmed that the JTAG port was enabled; this can be used to interact with the transmitter:

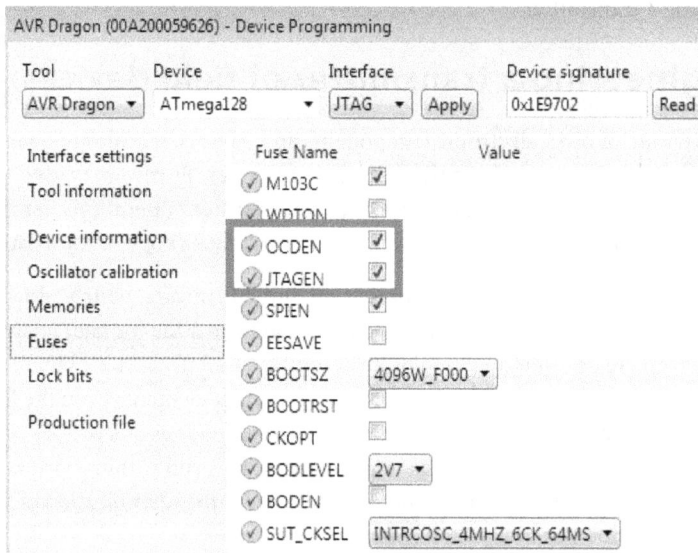

Figure 4.17 – JTAG port enabled

Finally, we will delve into the most sophisticated method – the bypass of a key switch.

Bypassing a key switch

Reid Wightman presented a method to bypass SIS key switches at the *S4* conference (`https://s4xevents.com/`). He showed how a malicious attacker could bypass the physical key switch lockout by gaining physical access to the SIS, reprogramming the main processor to bypass the locking mechanism, and then using the debug interface to take control of the system without requiring the use of the key switch.

The first step is to gain physical access to the SIS. Once access is obtained, a specialized debug cable is connected to the main processor on the SIS. This gives the malicious attacker access to the processor and enables them to read and write memory, as well as control any underlying processors on the SIS. The attacker then needs to reprogram and bypass the physical key switch lockout mechanism. This step is accomplished by writing a new operation code that allows the attacker to take control of the system without the use of the physical key.

The next step is to bypass the administrative security system. This is done by exploiting common vulnerabilities in the system's firmware. Once a vulnerability is found, the attacker has to write malicious code to overwrite the existing code and then re-flash the firmware. This will allow an attacker to gain administrative control of the system, and from here, they will be able to alter system settings, control functions, and disable security measures.

Finally, the malicious attacker will use the debug interface to gain complete control of the system. Often, debug interfaces are left unprotected, so an attacker can read and write memory, modify registers, and manipulate execution paths. By exploiting this, the attacker can gain full control of the SIS and effectively bypass the physical key switch lockout.

Overall, Reid Wightman's method of bypassing the physical key switch lockout provides an effective way for malicious attackers to gain control of an SIS. It requires physical access, the exploitation of vulnerabilities in the firmware, and the use of debug interfaces. However, if these steps are carried out successfully, an attacker is able to gain full control of the system, allowing them to access sensitive data or disable safety mechanisms intended to protect the system.

An SIS key switch consists of six modes:

- **Remote**: Used to disconnect the local workstation and enable a remote workstation to interface with the safety system
- **Run**: Used to enable the system to function in a supervised state, monitoring I/Os to ensure the process operates safely
- **Program**: Used to program and reconfigure the system
- **Stop**: Used to suspend the system and display diagnostic messages
- **Local**: Used to take back control of the local workstation from the remote
- **Monitor**: Used to observe and verify system responses without having control of the functions

The following photo represents a controller with a (black) key switch in run mode:

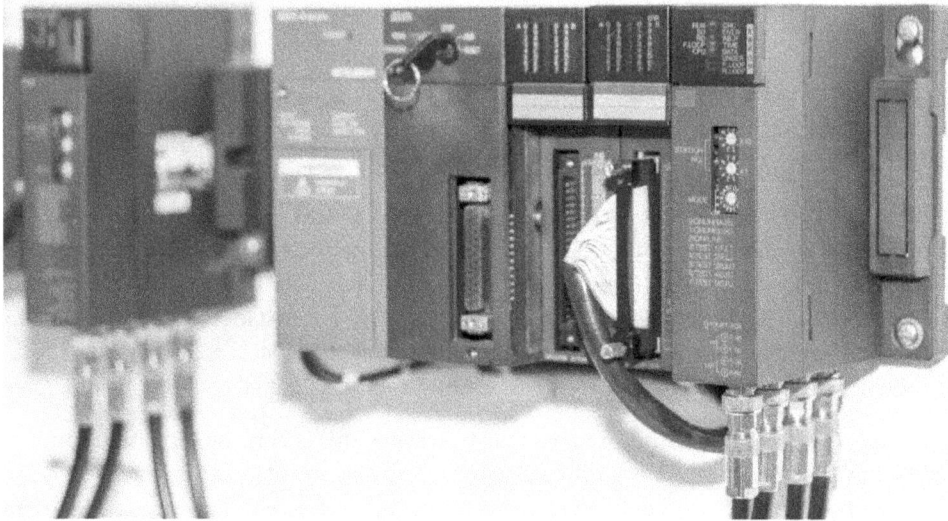

Figure 4.18 – Controller with key switch

Based on the various SIS attack vectors that can be targeted in this field, we will now look at an example of a safety controller attack with a focus on popular entry points.

Putting it all together

It's clear that threat actors now have a wide attack surface and, most likely, the time and motivation needed to study and fully prepare their attacks, as we have seen from previous incidents. The reality is that the majority of ICS environments are still operating with vulnerable installations – but this does not necessarily affect the risk profile of these systems since other safeguards are in place to mitigate cyber threats.

In the meantime, we all agree that if an attacker is able to gain access to these safety systems, they can modify the safety parameters and processes of the system, meaning the system may not be able to detect a dangerous hazardous condition and take corrective action, resulting in a disaster.

As we still have not witnessed such scenarios, this is a strong indication that the layered approach and mechanical protections deployed by the wider industry are still working.

> **Important note**
> Due to the confidential nature and constraints related to the purchase of SISs (only available for asset owners/operators) as well as the high costs associated with components, we decided to use a small safety PLC for some of the exercises, as this remains relatively affordable at present.

In the next section, we will conduct a lab exercise in the form of a safety PLC security assessment, including the tools and a step-by-step approach to evaluate the device's security.

Lab exercise – ReeR MOSAIC M1S safety PLC security assessment

The main aim of this lab exercise is to evaluate the cybersecurity measures implemented in the ReeR MOSAIC M1S safety PLC. This will involve assessing its vulnerability and conducting a mapping against the MITRE ICS framework.

For this lab, you will need the following:

- **Hardware**: ReeR MOSAIC M1S safety PLC (or other safety PLC): `https://www.reersafety.com/en/categories/safety-controllers-and-interfaces/`

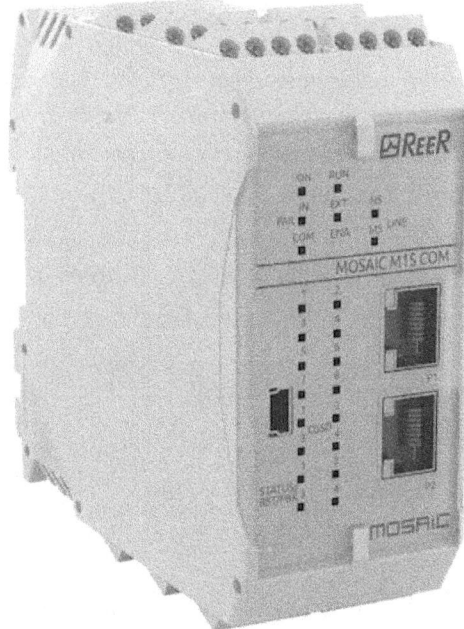

Figure 4.19 – ReeR MOSAIC M1S COM

- **Software**:

  - Download the MOSAIC M1S COM configuration file from here: `https://www.reersafety.com/en/software-2/` (Registration required)
  - Windows, Mac, or Linux machine

- **Wireshark**: Network protocol analyzer that captures and displays data traveling back and forth on a network.

- Metasploit or other exploitation framework

This lab is meant to give students or security practitioners an opportunity to conduct a hands-on cybersecurity assessment. Performing these steps will help you develop analytical skills for detecting vulnerabilities and preventing and providing mitigating countermeasures to improve product security.

> **Disclaimer**
>
> This exercise should ideally be supervised by an experienced instructor, and any testing or probing should be carried out within the confines of an isolated and secure laboratory environment. Caution should be exercised not to disrupt actual operational environments or violate any laws or ethics guidelines in the process.

The following steps will be carried out as part of the lab exercise:

1. **Study the documentation**: Review the ReeR MOSAIC M1S safety PLC documentation thoroughly. Pay special attention to areas pertaining to the device's cybersecurity features.

2. **Baseline setting**: Configure the ReeR MOSAIC M1S safety PLC with pre-determined settings to establish a baseline for the assessment.

3. **Network traffic analysis**: Using Wireshark, analyze ongoing network traffic involving the ReeR MOSAIC M1S safety PLC. Look for signs of possible intrusion or data leakage.

4. **Vulnerability scanning**: Conduct a thorough vulnerability scan of the ReeR MOSAIC M1S safety PLC using Nessus.

5. **Data analysis**: Analyze your findings using Python. Detect anomalies and vulnerabilities while also evaluating the severity of each one.

The lab will be initiated and conducted in the following steps:

1. **Threat modeling**:

    I.   Identify potential threats to the safety PLC, including deliberate attack vectors and potential entry points.

    II.   Evaluate vulnerabilities in the safety PLC, such as flaws in network communications, hardware, software, and operational processes.

    Next is a sample threat modeling analysis based on the identified attack vectors and their alignment with relevant sample MITRE ICS tactics and techniques:

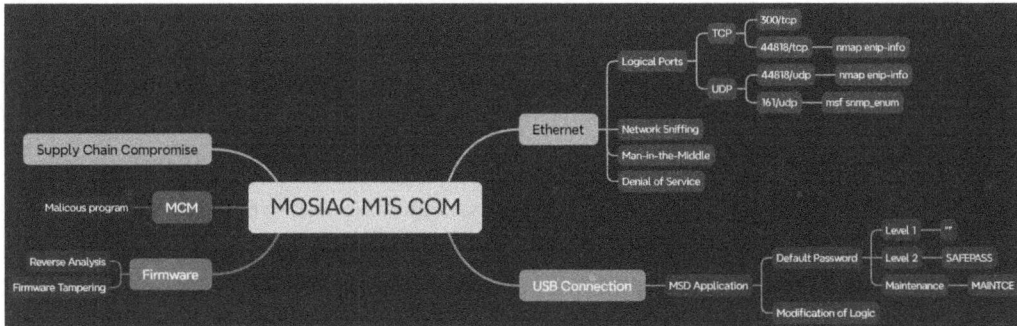

Figure 4.20 – Example of identified attack vectors

Based on the supported protocols (Ethernet/IP, MODBUS/TCP, PROFINET, EtherCAT) and hardware interfaces (USB, LAN) of the MOSAIC M1S COM module, several possible attack vectors can be identified. Aligning these vectors with the MITRE ICS framework helps in understanding potential threats and planning for appropriate defenses.

2. **Hardware and software profiling**:

I.   **Examine the USB interface**: Demonstrate how the mini-USB 2.0 connector enables direct connection to a personal computer for programming and configuration via the **MOSAIC Safety Designer (MSD)** software interface.

II. **Explore the LAN interface (RJ45)**: Show how the RJ45 ports facilitate TCP/IP LAN network connectivity, allowing the exchange of process data across various Fieldbus protocols and remote configuration of the device.

These are some findings for hardware analysis:

• **USB interface**:

A mini-USB 2.0 (USB-C) connector on the MOSAIC M1S COM module allows for a direct connection to a personal computer. Using the graphical MSD software interface, which allows the creation of complicated logic and safety functions, this interface makes it easier to program and configure the device.

• **LAN interface (RJ45)**:

For TCP/IP LAN network connectivity, the device additionally has two RJ45 ports. This facilitates the interchange of process data across certain Fieldbus protocols and permits remote configuration of the MOSAIC M1S COM module over a network connection. For Fieldbus connectivity, the device supports Ethernet/IP, MODBUS/TCP, PROFINET, and EtherCAT; DHCP and DNS support varies depending on the protocol.

3. **Protocol support**:

    I.   **Investigate Ethernet/IP**: Discuss its role in industrial automation, offering real-time data gathering and management.

    II. **Explore MODBUS/TCP**: Explain how it facilitates communication over TCP/IP networks.

    III. **Explore PROFINET**: Discuss its role in data communication over **Industrial Ethernet (IE)** for equipment management within industrial systems.

    IV. **Explore EtherCAT**: Highlight its use for real-time automation tasks in industrial settings.

    With Wireshark, we can capture network traffic:

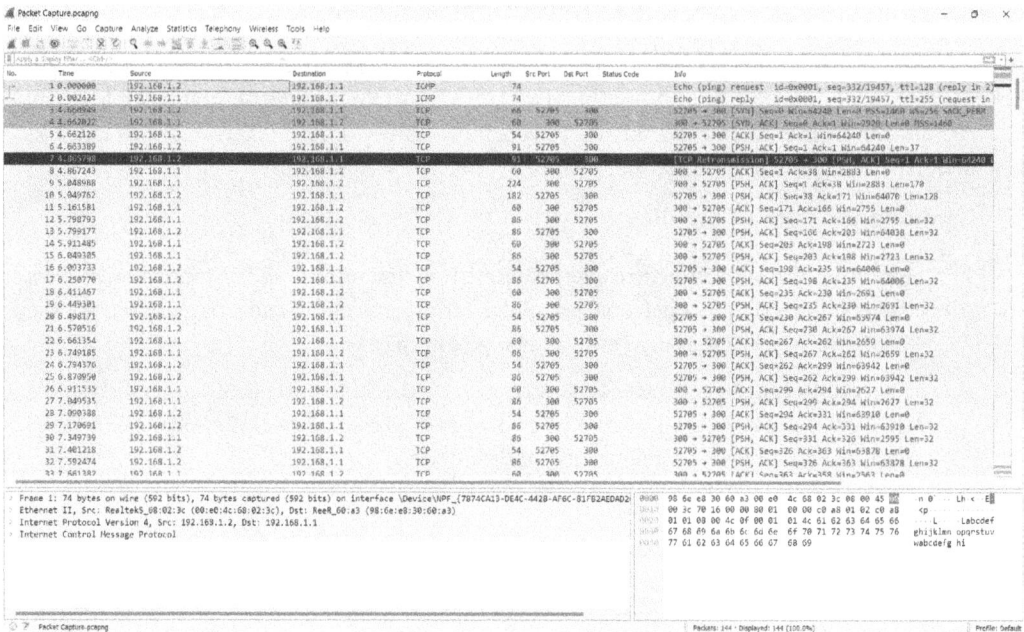

Figure 4.21 – Wireshark packet capture

4. **Sample tactics, techniques, and procedures (TTPs)**:

    I.   Present several TTPs associated with cyber threats to the safety PLC, such as network eavesdropping, spoofing and session hijacking, **Denial-of-Service (DoS)** attacks, firmware tampering, and so on.

    II. Demonstrate how each TTP could be executed and its potential impact on the safety PLC.

    Several TTPs could be identified and executed for our safety PLC. Next is a sample set of these TTPs:

    i.   **Network eavesdropping attack** (**Tactics**: *T1 - Initial Access*, **Technique**: *T0819 - Exploit Public-Facing Application*)

- **Attack technique**: Utilizing packet sniffers and network monitoring tools to capture unencrypted traffic over the network, attackers identify sensitive data, credentials, or any relevant information that can be utilized for further attacks.

- **Tools**: Wireshark, `tcpdump`, Ettercap.

- **Strategy**: Position the attack at strategic points (for example, network hubs or gateways) to maximize data interception, possibly using ARP spoofing to redirect traffic.

ii. **Spoofing and session hijacking** (**Tactics**: *T1 - Initial Access*, **Technique**: *T817 - Drive-by Compromise*)

- **Attack technique**: Impersonating devices or sessions to intercept or inject malicious data into the communication stream. This could involve ARP spoofing, DNS poisoning, or session token hijacking.

- **Tools**: Cain & Abel, BetterCAP, Metasploit.

- **Strategy**: Target communication sessions that are not encrypted or are poorly authenticated, then take over the session to execute unauthorized commands or extract data.

iii. **DoS attacks** (**Tactics**: *T10 - Inhibit Response Function*, **Technique**: *T0803 - Block Command Message*)

- **Attack technique**: Overwhelming the device or network with excessive traffic or requests, exploiting specific vulnerabilities to crash systems or exhaust resources.

- **Tools**: **Low Orbit Ion Cannon (LOIC)**, **High Orbit Ion Cannon (HOIC)**, stress-testing tools such as `hping3`.

- **Strategy**: Identify the weakest link in the network or application layer and target it with traffic that is difficult for the system to process, leveraging botnets for amplification if necessary.

iv. **Firmware tampering** (**Tactics**: *T3 - Persistence*, **Technique**: *T0889 - Modify Program*)

- **Attack technique**: Modifying the device's firmware by exploiting update mechanisms, physical access, or vulnerabilities in the firmware itself. The goal is to embed malicious functionality or create backdoors.

- **Tools**: Firmware modification kits, JTAG debuggers.

- **Strategy**: Research publicly disclosed vulnerabilities or use reverse engineering to find new ones, then craft malicious firmware updates or use physical interfaces to flash the device.

v. **MitM attacks** (**Tactics**: *T8 - Collection*, **Technique**: *T0830 - Adversary in the Middle*)

- **Attack technique**: Intercepting and potentially altering communications between two parties without their knowledge. This can be done by compromising the network, using ARP spoofing, or exploiting weak encryption.

- **Tools**: `mitmproxy`, Ettercap, BetterCAP.

- **Strategy**: Gain access to the communication channel, decrypt the traffic, and passively monitor or actively manipulate the data being transmitted.

vi. **Universal Serial Bus (USB)/ Multi-Chip Module (MCM)-based malware introduction**
    Tactics: *T1 - Initial Access*, **Technique**: *T0847 - Replication Through Removable Media*

- **Attack technique**: Using physical access to insert USB devices that contain malicious software into the SIS device. The malware could be designed to exploit vulnerabilities, establish backdoors, or disrupt operations.

- **Tools**: Rubber Ducky, Bash Bunny.

- **Strategy**: Create or use existing malware that auto-executes upon USB insertion, possibly employing social engineering to trick personnel into plugging in the infected USB device.

vii. **Configuration/parameter tampering** (**Tactics**: *T7 - Lateral Movement*, **Technique**: *T0843 - Program Download*)

- **Attack technique**: Altering device configurations or operational parameters to disrupt functionality or disable safety features. Achieved through unauthorized interface access, exploiting services for configuration changes, or malicious updates.

- **Tools**: Network scanning tools such as Nmap and exploitation frameworks such as Metasploit.

- **Strategy**: Identify accessible configuration interfaces using reconnaissance. Exploit authentication or validation weaknesses to access and modify configurations, aiming to subtly compromise system integrity or disable safety mechanisms without detection.

viii. **Supply chain compromise** (**Tactics**: *T1 - Initial Access*, **Technique**: *T0862 - Supply Chain Compromise*)

- **Attack technique**: Compromising the integrity of software or hardware components at any point in the supply chain. This could involve inserting malicious code into the software before it's delivered to the end user or tampering with hardware components to introduce vulnerabilities.

- **Tools**: Malware, compromised firmware, or hardware modification tools.

- **Strategy**: Identify targets within the supply chain who have less secure practices, then infiltrate their systems to insert malicious components into the products or software. These compromised elements could then provide backdoor access or disrupt operations upon deployment in the target environment.

ix. **Common port abuse** (**Tactics**: *T9 - Command and Control*, **Technique**: *T0885 - Commonly Used Port*)

- **Attack technique**: Attackers target network ports (for example, TCP/80, HTTPS TCP/443, TCP/102, TCP/502, UDP/161) to identify and exploit vulnerabilities in services running, gain unauthorized access, or perform reconnaissance activities. Network ports are often targeted because they are likely to be open and accessible to attackers.

- **Tools**: Nmap, Zenmap, Metasploit

- **Strategy**: Perform network scans on the target device to identify used TCP/UDP ports. This could involve using Nmap script scanning engine to capture banners, perform vulnerability scanning, and gather further information.

These offensive venues highlight the methodologies attackers might deploy to exploit vulnerabilities in devices. Understanding these can help in developing more robust defenses against potential cyber threats. Aligning these attack vectors with the MITRE ICS framework not only highlights potential threats but also guides the development of targeted defense strategies to protect the MOSAIC M1S COM module and associated systems from cyber threats.

Based on the threat modeling activity, the following techniques can be selected, and adequate tests can be performed:

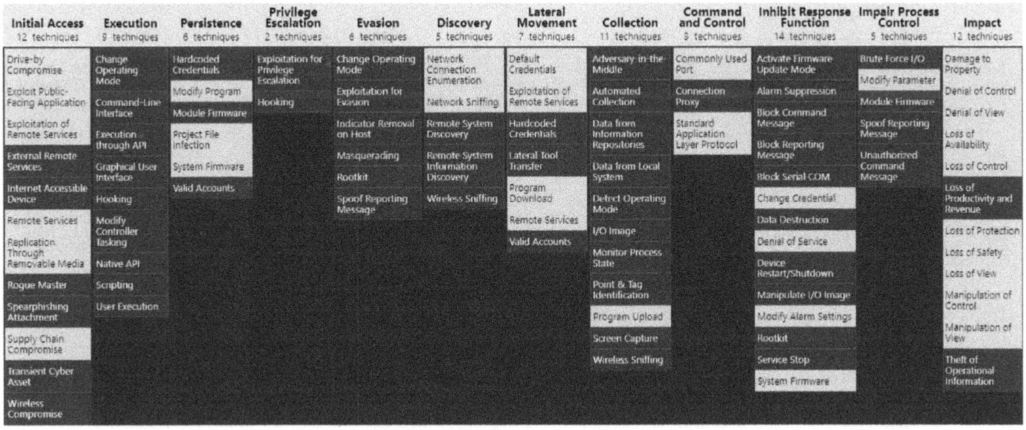

Figure 4.22 – MITRE ICS mapping

5. **Tools and techniques**:

A. Conduct Nmap scans to identify ports of interest and vulnerabilities.

B. Utilize Metasploit to gather information about the device configuration.

C. Capture and analyze network traffic using Wireshark to identify potential threats.

D. Conduct **Open Source Intelligence** (**OSINT**) analysis to gather additional information about the device.

6.   Next, you'll see an example of the output from the four activities (*A*, *B*, *C* and *D*):

- Nmap TCP scan

Execute the following Nmap command, and you will get the output as shown in *Figure 4.23*:

```
nmap -sS -sV -p - -v 192.168.1.1
```

```
Nmap scan report for 192.168.1.1
Host is up (0.0040s latency).
Not shown: 65533 closed tcp ports (reset)
PORT       STATE SERVICE       VERSION
300/tcp    open  unknown
44818/tcp  open  EtherNet-IP-2
1 service unrecognized despite returning data. If you know the service/versio
please submit the following fingerprint at https://nmap.org/cgi-bin/submit.cg
new-service :
SF-Port300-TCP:V=7.92%I=7%D=3/14%Time=65F33237%P=i686-pc-windows-windows%r
SF:(GetRequest,20,"w\$\xf9\xd7\x16o\xc3\xaa\x14\xf5q\x0e3\xbc\xa5\xd4\x13v
SF:\$@\xd2Hf9\x8e\x90\xd3\x14\x1eD\xfb\xef")%r(HTTPOptions,20,"\x03\x08\0\
SF:x0e6AES_START_FAI\nx\0\0\0\0\0\0\0\0\0\0\0")%r(RTSPRequest,20,"\$#\xc
SF:3C\xbe\xc50\x8f5\x11\xb1\xfei\x85\x9ff\x15\x0c\xc8\xe3\xfcV\xe9\x1e\xba
SF:\xb1\x9bGd5\xe80")%r(RPCCheck,20,"\xe3R\xec\x08\$\x99\x85R\xfe\0\xa4\xd
SF:7\r\x06\x97\xc1\x7f\x0c\xb2\xd6k\xac\xdb\xc6C\xa0cy\xea\xb2\xec\xdf")%r
SF:(DNSVersionBindReqTCP,20,"\xa3\x80\x16\xcc\x8b1\xd9\x15\xc6\xef\x98\xaf
SF:\xb2\x88\x8f\x1c\x8b\xe8P\[\xad\$k\x81\xa0\xbf\xf9\xde\x16\xab\xbaB")%r
SF:(DNSStatusRequestTCP,20,"\x03\x08\0\x0e6AES_START_FAI\nx\0\0\0\0\0\0\0\
SF:0\0\0\0\0")%r(Help,20,"\x03\x08\0\x0e6AES_START_FAI\nx\0\0\0\0\0\0\0\
SF:0\0\0\0")%r(SSLSessionReq,20,"\xe8\xdf9\xb4>\xefw\x80c6\xbd\xa4\x1b\xc3
SF:\xd9\x20\xaa8\xbeW\xa4\xa8\xc9\x90\x15P\x01Co\x84-\xde")%r(TerminalServ
SF:erCookie,20,"\xa8\rbx\xa4\xc3\xcbB,%\xb1\|\xc0D\xd1{k\xcb\x8ed\xba\xce\
SF:x9a\x86f\xb7\xdc\xc4\x15n\xbb\xc9")%r(TLSSessionReq,20,"\x03\x08\0\x0e6
SF:AES_START_FAI\nx\0\0\0\0\0\0\0\0\0\0\0")%r(Kerberos,20,"\x03\x08\0\x0
SF:e6AES_START_FAI\nx\0\0\0\0\0\0\0\0\0\0\0")%r(SMBProgNeg,20,"!\xc8\xba
SF:U\xb1\x14\xe5\xda\xb6\x1d\xfa\xd9\x9eEzF\xa0\xc5\xd8\x06X\xc1B%\xb7\x16
SF:m\xac\xac/\xee\xf2")%r(X11Probe,20,"\xe1\xf6\xe3\x19\x17\xe79\x9d\x7f\x
SF:0c\xee\xb1C\xc7q\xa1w\x88\x87o\xbe\x88\0\x17\xd1D\xe8:\xbbGj}")%r(FourO
SF:hFourRequest,20,"\x03\x08\0\x0e6AES_START_FAI\nx\0\0\0\0\0\0\0\0\0\0\
SF:0")%r(LPDString,20,"\x03\x08\0\x0e6AES_START_FAI\nx\0\0\0\0\0\0\0\0\0\0
SF:\0\0")%r(LDAPSearchReq,20,"\x03\x08\0\x0e6AES_START_FAI\nx\0\0\0\0\0\
SF:0\0\0\0\0\0")%r(LDAPBindReq,20,"\x03\x08\0\x0e6AES_START_FAI\nx\0\0\0\0
SF:\0\0\0\0\0\0\0")%r(SIPOptions,20,"\x03\x08\0\x0e6AES_START_FAI\nx\0\0
SF:\0\0\0\0\0\0\0\0")%r(LANDesk-RC,20,"\x03\x08\0\x0e6AES_START_FAI\nx
SF:\0\0\0\0\0\0\0\0\0\0\0")%r(TerminalServer,20,"\x03\x08\0\x0e6AES_STAR
SF:T_FAI\nx\0\0\0\0\0\0\0\0\0\0\0")%r(NCP,20,"i\xe4\xd5\xd9:\xa2\x87\xa8
SF:\xf4/\xe4\xb3\$\xef%\xb4\xf9\x92\x87\x05\x1d\xc5\xf3\xbaU\xb4R\x81tQW>"
SF:)%r(NotesRPC,20,"\x03\x08\0\x0e6AES_START_FAI\nx\0\0\0\0\0\0\0\0\0\
SF:0");
MAC Address: 98:6E:E8:30:60:A3 (Ieee Registration Authority)
```

Figure 4.23 – Nmap TCP scan output

We identified two ports (`300/tcp` and `44818/tcp`) of interest that will help us to define how can interact with the device.

Furthermore, the UDP scan revealed some more ports.

Run the following UDP scan command to get the output shown in *Figure 4.24*:

```
nmap -sU -p - -v 192.168.1.1
```

```
Nmap scan report for 192.168.1.1
Host is up (0.0060s latency).
Not shown: 65531 closed udp ports (port-unreach)
PORT        STATE           SERVICE
161/udp     open|filtered snmp
2222/udp    open|filtered msantipiracy
25383/udp   open|filtered unknown
44818/udp   open|filtered EtherNetIP-2
MAC Address: 98:6E:E8:30:60:A3 (Ieee Registration Authority)
```

Figure 4.24 – Nmap UDP scan output

Let's now enumerate the device on these ports:

```
nmap -sS -p 44818 -v --script enip-info 192.168.1.1
```

```
Nmap scan report for 192.168.1.1
Host is up (0.0011s latency).

PORT        STATE SERVICE
44818/tcp open   EtherNet-IP-2
| enip-info:
|    type: Communications Adapter (12)
|    vendor: Hilscher GmbH (283)
|    productName: M1SCOM
|    serialNumber: 0x00000c26
|    productCode: 5
|    revision: 1.2
|    status: 0x0034
|    state: 0x03
|_   deviceIp: 192.168.1.1
MAC Address: 98:6E:E8:30:60:A3 (Ieee Registration Authority)
```

Figure 4.25 – Port enumeration result

Now, let's enumerate port 161 to gather more information about the target device:

```
nmap -sU -p 161 -v --script snmp-brute 192.168.1.1
```

```
Nmap scan report for 192.168.1.1
Host is up (0.0018s latency).

PORT      STATE SERVICE VERSION
161/udp open   snmp    SNMPv1 server (public)
MAC Address: 98:6E:E8:30:60:A3 (Ieee Registration Authority)
```

Figure 4.26 – SNMP scan result

- Metasploit SNMP scan

The Metasploit scan could provide good information about the device configuration, as depicted in the following screenshot:

```
msf6 auxiliary(scanner/snmp/snmp_enum) > run

[+] 192.168.1.1, Connected.

[*] System information:

Host IP                             : 192.168.1.1
Hostname                            : -
Description                         : -
Contact                             : -
Location                            : -
Uptime snmp                         : -
Uptime system                       : -
System date                         : -

[*] Network interfaces:

Interface                           : [ up ] my switch
Id                                  : 1
Mac Address                         : 98:6e:e8:30:60:a3
Type                                : ethernet-csmacd
Speed                               : 100 Mbps
MTU                                 : 1500
In octets                           : 9207497
Out octets                          : 8149408

Interface                           : [ up ] my switch Port 1
Id                                  : 2
Mac Address                         : 98:6e:e8:30:60:a4
Type                                : ethernet-csmacd
Speed                               : 100 Mbps
MTU                                 : 1500
In octets                           : 9286100
Out octets                          : 8548979

Interface                           : [ up ] my switch Port 2
Id                                  : 3
Mac Address                         : 98:6e:e8:30:60:a5
Type                                : ethernet-csmacd
Speed                               : 100 Mbps
MTU                                 : 1500
In octets                           : 0
Out octets                          : 0
```

Figure 4.27 – Metasploit scan result

During the SNMP scan, a variety of valuable information can be captured, including MAC addresses, contact details, hostnames, and more.

On OSINT, plenty of information can be found, including passwords and manuals:

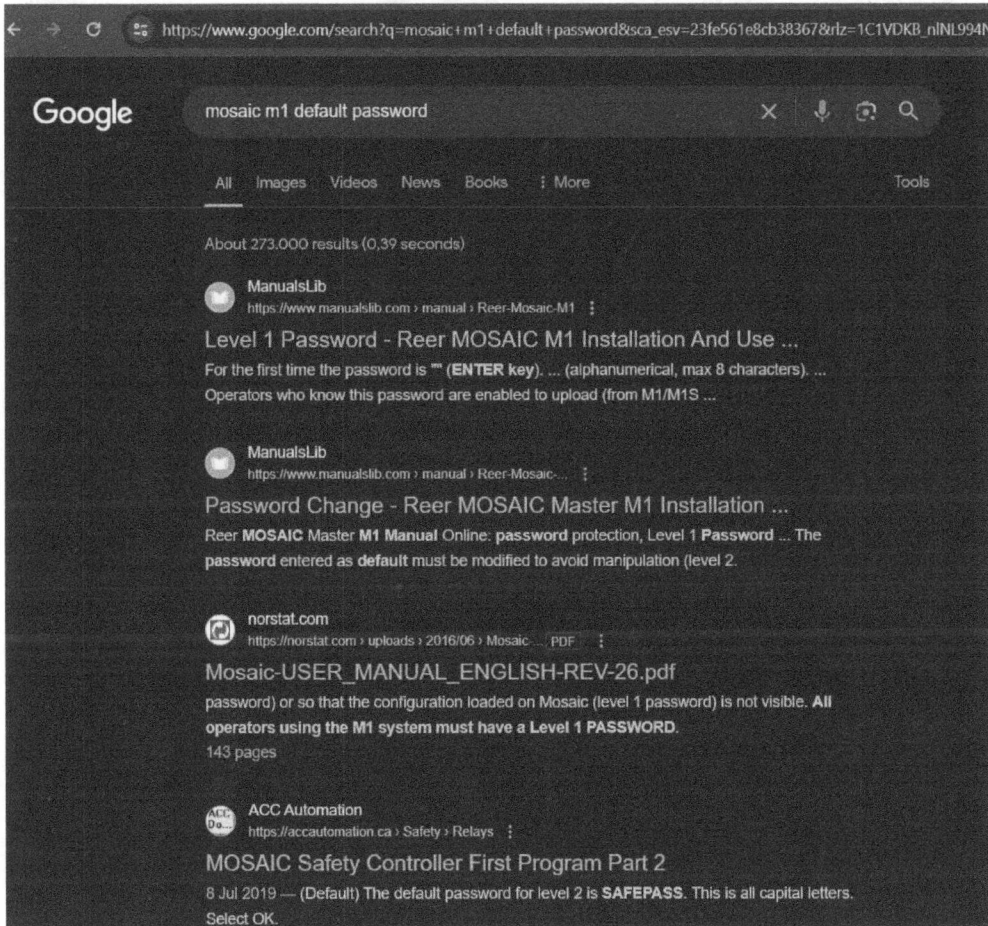

Figure 4.28 – OSINT analysis

In addition to the preceding techniques, it's possible to interact and configure the safety PLC via the MSD application:

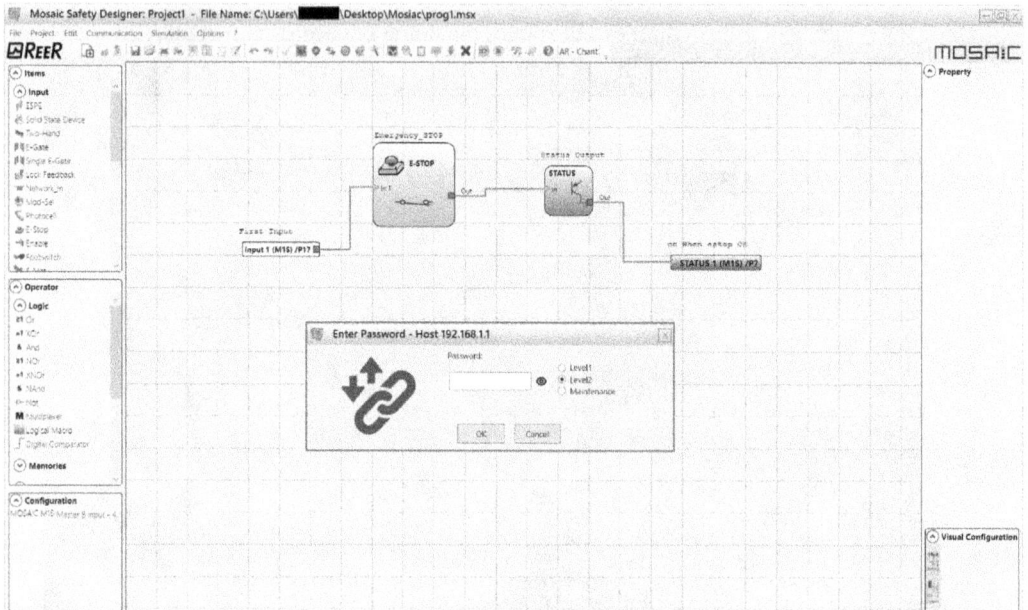

Figure 4.29 – MSD authentication

The access level is based on three levels of users – administrator (Level 2), a basic user with no privileges, and a maintenance account:

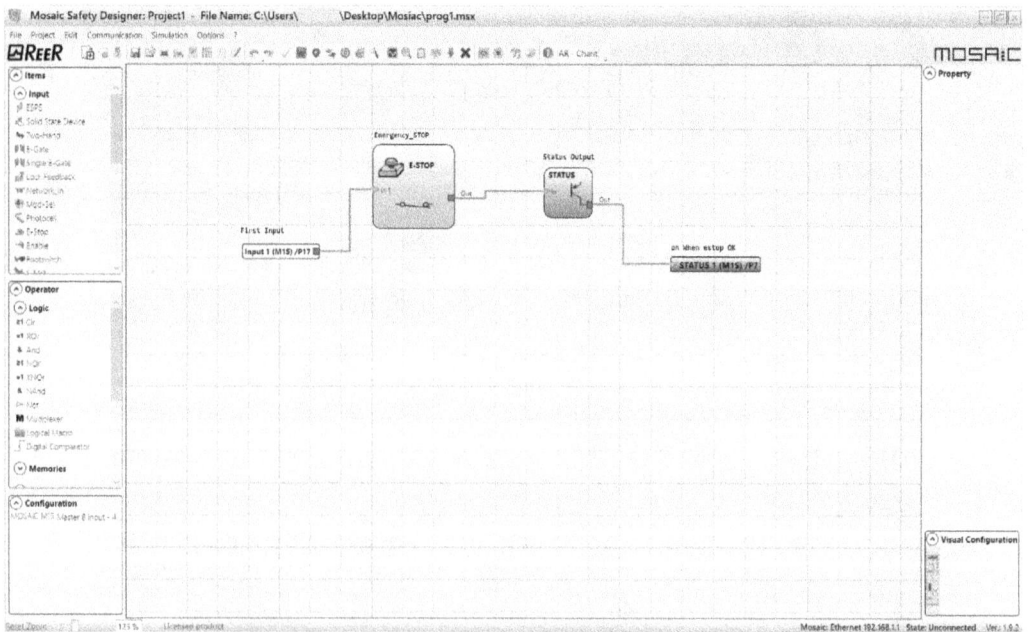

Figure 4.30 – MSD project

Of course, with this access, you can configure the safety PLC and make changes to the logic as required.

In this lab, we established a foundational framework for assessing the security of a safety PLC. The evaluation included performing a port mapping, vulnerability scan, and gathering OSINT information to determine whether the product meets minimum security baselines.

Summary

In this chapter, we examined commonly used attack techniques in the field of SISs, highlighting various aspects and evidence associated with previous attacks and incidents. These findings serve as valuable resources for implementing effective countermeasures and conducting further security research in this domain.

The next chapter will dive into security measures that can be implemented to safeguard ICSs and SISs from malicious entities, emphasizing the importance of people, processes, and technology in mitigating deliberate or accidental threats.

Further reading

- Lockheed Martin, the Cyber Kill Chain® framework: `https://www.lockheedmartin.com/en-us/capabilities/cyber/cyber-kill-chain.html`

- *International Society of Automation, InTech, January/February 2020*: `https://www.isa.org/getmedia/f0e65214-e8d0-44b5-9ff5-cc31557b2aff/InTech-January-February-2020.pdf`

- *A survey of offensive security research on PLCs*: `https://iopscience.iop.org/article/10.1088/1742-6596/1976/1/012025/pdf`

- *TRITON: The First ICS Cyber Attack on Safety Instrument Systems – Understanding the Malware, Its Communications and Its OT Payload: Alessandro Di Pinto, Younes Dragoni, Andrea Carcano, Black Hat USA 2018 - Research Paper*: `https://icscsi.org/library/Documents/Cyber_Events/Nozomi%20-%20TRITON%20-%20The%20First%20SIS%20Cyberattack.pdf`

- *Malware attacking commonly used in Industrial Control Systems (ICS) Triconex Safety Instrumented System (SIS) controllers*: `https://malpedia.caad.fkie.fraunhofer.de/details/win.triton`

- *A threat model method for ICS malware: the TRISIS case*: `https://dl.acm.org/doi/abs/10.1145/3457388.3458868`

- *NAMUR, WG 4.5 Functional Safety*: `https://www.namur.net/en/work-areas-and-project-groups/wa-4-operation-support-and-maintenance/wg-45-functional-safety.html`

- *Dragos, TRISIS: Analyzing Safety System Targeting Malware*:

  `https://www.dragos.com/resource/trisis-analyzing-safety-system-targeting-malware/`

- Schneider Electric downloads: `https://www.se.com/ww/en/download/document/SEVD-2017-347-01/`

- Metasploit framework: `https://www.metasploit.com/download`

- Nessus scanner: `https://www.tenable.com`

- Nmap: `https://nmap.org`

5
Securing Safety Instrumented Systems

In the previous chapter, we explored the multifaceted attack landscape of **Industrial Control Systems** (**ICSs**), examining how adversaries may exploit systematic weaknesses, inherent vulnerabilities, and insecure-by-design issues. We cast a light on the myriad ways that these critical systems, often designed with (functional) safety prioritization in lieu of cybersecurity, can be unwilling conduits for cyber threats. With this foundation laid, we advance our journey in this chapter by discussing a strategic array of cybersecurity countermeasures tailored to safeguard SISs against emerging cyber attacks and to enhance their overall risk profile.

This chapter succinctly presents the core principles of ICS cybersecurity essential for protecting **Safety Instrumented Systems** (**SISs**) across their lifecycle. We explore critical security controls and practices – encompassing cyber, physical, managerial, and engineering aspects – to maintain the integrity and security of SISs. We begin by discussing how integrating security from the design stage can preemptively address threats to SISs, emphasizing the importance of strategic planning, appropriate material selection, and robust backup and recovery processes to fortify defenses against cyber attacks. The chapter also delves into **Cybersecurity Management Systems** (**CSMSs**) and the need for structured, adaptable strategies to manage cyber risks effectively.

Further, we underscore the necessity of a corporate framework that merges IT with **Operational Technology** (**OT**), fostering inclusive governance for cyber risk management and promoting cybersecurity as a shared organizational responsibility. We then detail methods for testing and evaluating SIS security, including vulnerability assessments and simulated attack exercises, to identify and mitigate potential weaknesses.

Finally, we highlight the ongoing commitment required in the operation and maintenance phases to preserve effective cybersecurity measures, ensuring SISs remain a benchmark of operational safety and cyber resilience.

This chapter will address the following topics:

- Security design and engineering
- CSMSs
- **Governance Operating Model (GOM)**
- Operation and cybersecurity maintenance

Security design and engineering

The cybersecurity of SISs is paramount throughout their lifecycle, from design and engineering to operation and maintenance. The role of design and engineering in ICS cybersecurity can't be overstated. It is the fusion of advanced security features, meticulous engineering, and responsible operations that creates safe and reliable assets. This requires not just a one-time effort but a commitment to continuous excellence and vigilance throughout the entire lifecycle of the system to protect against the incessant evolution of cyber threats in the dynamic landscape of ICSs.

In this section, we will discuss the **secure-by-design** principle as a promising model to overcome the challenges of legacy installation and insecure products.

The secure-by-design principle

In recent years, the secure-by-design concept has gained significant traction in the technology industry, garnering attention and accolades for its effective approach to cybersecurity. However, despite its widespread recognition and proven benefits, many suppliers in the market struggle to fully embrace and implement this principle due to cost constraints and the relentless push for quick product releases. This has created a disparity between the concept's potential and its actual industry adoption.

It is a principle that has evolved over time in the field of cybersecurity and has been advocated by various organizations and industry experts. Some early contributions to the secure-by-design concept can be traced back to the work of security researchers at the **National Security Agency** (**NSA**) and the **Information Sciences Institute** (**ISI**) in the 1980s and 1990s. However, the term *secure by design* has gained more prominence in recent years with the rise of initiatives such as the **Industrial Internet of Things Consortium** (**IIC**) and the European Union's **General Data Protection Regulation** (**GDPR**), which emphasizes the importance of incorporating security measures into the design of information systems as well as industrial internet systems.

The secure-by-design concept encompasses both hardware and software components of a product, providing comprehensive protection against potential cyber threats. While this approach has been widely adopted in traditional IT systems, its implementation in crucial ICS and SIS products is still in its early stages. This is mainly due to the complex nature of these systems and the inherent focus on functional safety and fail-safe measures, making cybersecurity an emerging aspect of their design and development. As a result, the incorporation of secure-by-design principles in these systems is still a work in progress and requires further attention and investment for maximum effectiveness.

Secure by design is an approach that focuses on developing systems with security in mind from the very beginning. This approach is particularly important when it comes to ICSs, which are responsible for controlling and monitoring critical processes in various process industries. Let's explore this in detail to understand the specification of this concept and how this can be adopted for product cybersecurity as well as asset life cycles.

Managing the ICS cybersecurity lifecycle

ICSs are typically deployed with an anticipation of extended operational life. However, continuous advancements, significant overhauls, customizations, and business consolidation requirements imply that several ICS-associated projects often arise during both the developmental and operational stages. These projects might potentially drive security concerns. Therefore, it becomes essential to maintain security over the complete ICS lifecycle, which encapsulates four primary stages: *design*, *build*, *operation*, and *decommissioning* (as depicted in *Figure 5.1*). After conducting a risk evaluation, any running or prospective projects that could possibly impact ICS security should employ a strategy that not only incorporates but also sustains security from the initial stages. The assurance of fulfilling these requirements should be periodically verified across the asset lifecycle. Any brand-new system on a "greenfield" site should include the security requirements in the process from the procurement, design, and execution stages.

The ICS's subsequent operation should ensure the efficiency of this embedded security during the rest of the system's service span. In the meantime, incorporating security safeguards into ICSs can be a challenging and expensive process if the systems are already established and functioning. Moreover, appending security elements to an already operational ICS or late in its development phase tends to be less successful compared to embedding them at the onset. Addressing security threats by introducing protective actions into the ICS during its initial stages of the lifecycle is typically more productive, helps avoid deviations, is generally less expensive, and can often act as a facilitator for business development.

In order to manage the ICS lifecycle within an organization, the following stages are described in *Figure 5.1* and in accordance with industry best practices:

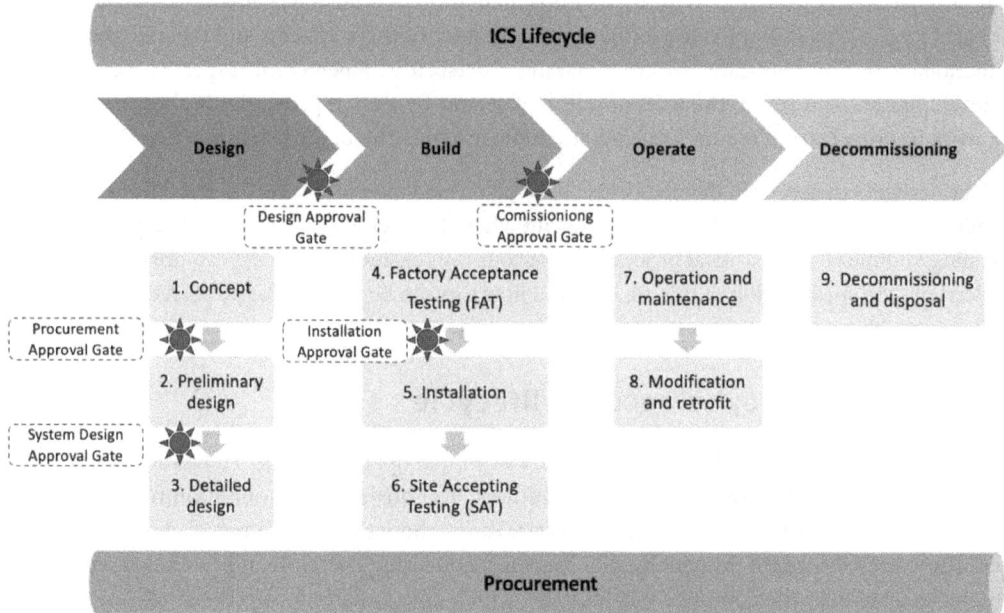

Figure 5.1 – Stage of ICS lifecycle

This process is often segmented into four distinguished pillars:

- **Design**
- **Build**
- **Operate**
- **Decommissioning**

Prior to the design phase, the conceptual planning phase is carried out to ensure project readiness. This includes initial risk analysis as part of a feasibility study, which is one of the actions for the conceptual phase.

Each stage consists of well-defined activities and tasks to ensure all risks or weaknesses in systems are identified and addressed accordingly in the following sections.

Design

During the design phase, the foundation for cybersecurity is set. This phase aligns closely with the principles of proactive protection where the emphasis is on selecting components and systems that inherently possess strong security features. It is in this phase that threat modeling should be performed to anticipate possible attack vectors and to comprehend the risk landscape specific to the operational environment. By involving cybersecurity specialists early on in the vendor selection and procurement process, organizations can ensure that security considerations take center stage alongside functional requirements. The goal is to ensure that vendors being selected are cognizant of cybersecurity threats and equipped to handle them, thereby reducing the risk of introducing vulnerabilities into the control system environment right from the start.

During the design phase, strategic decisions are made with long-term security in mind. Security requirements are established, suppliers are scrutinized for their cybersecurity capabilities, and product offerings are evaluated against security benchmarks. This is where commitment to secure by design is firmly expressed by insisting on security controls:

- **Preliminary design**: As the project transitions into the design phase, the functional requirements that were established are transformed into detailed system specifications and security controls are specified in detail. This phase should incorporate a security risk assessment to pinpoint anticipated threats and vulnerabilities specific to the ICS, thereby dictating security features that must be engineered into the system. Establishing strong, unambiguous requirements for cybersecurity during this phase is critical for guiding the subsequent design and execution phases.

  Prior to initiating a detailed design, the system design will require approval.

- **Detailed design**: The **design and execute** phase involves the creation of detailed system architecture and the rollout of security protocols – both hardware and software. It's crucial that the design incorporates layers of security mechanisms, from physical access controls to network security and application-level safeguards. These might include **Role-Based Access Control (RBAC)** systems, firewalls, and **Intrusion Detection Systems (IDSs)**.

Completion of the detailed design sets the scene for the build (or execute) phase in which the system will be developed.

Build

Moving into the build phase, design and engineering practices come to the forefront. It's here that security needs to be embedded into the system architecture. System and network designers and engineers are tasked with integrating cybersecurity controls identified during the procurement phase into the actual development of the ICS and SIS. Implementing secure coding practices, ensuring proper component testing, and maintaining an exhaustive inventory of all devices and software dependencies are only the basics.

Engineers must also establish thorough communication protocols, encryption standards for data transmissions, and robust authentication mechanisms to thwart unauthorized access. Additionally, it's imperative that the SIS is designed with scalability and flexibility in mind, enabling it to adapt as threat landscapes evolve and new vulnerabilities are discovered.

Building with modularity allows individual components to be updated or replaced securely and efficiently. Rigorous testing, including penetration testing and technical security audits, should be conducted at this stage to assess the system's defense mechanisms. Execution with these principles enhances not only the security posture but also the reliability and longevity of ICS assets.

Factory Acceptance Testing (FAT), installation, and **Site Acceptance Testing (SAT)** are crucial parts of the ICS lifecycle. These tests provide a means to validate and confirm the correct operation of the system before it goes online and begins its operational life. Let's understand them:

- **FAT**: FAT is conducted at the vendor's site after the system is built but before it is shipped to the client's site. The purpose is to ensure that the system meets the agreed specifications and functions correctly as per the design. This can include checks on hardware components, software configurations, and ICS functionalities.

- **Installation**: After the system has passed the FAT and is shipped to the user, the next step is installation. This involves physically installing the equipment in its operational environment, which includes not just setting up the hardware components but also installing software, setting up network connections, implementing security measures, and so on. It is important to carefully plan and execute installation to avoid operational failures, safety risks, and costly downtime.

- **SAT**: SAT takes place at the installation site after FAT and installation. It serves to verify that the system operates as intended within its operational environment, confirming that the system interacts correctly with other systems and within the actual infrastructure it will be used.

    A successful SAT confirms that the system will operate as expected once it goes into operation.

All these tests – FAT, installation, and SAT – are vital to ensure a smooth transition to operation for any ICS. It allows for the early identification and resolution of problems, which helps in avoiding costly and dangerous failures once the system is live. In the next section, we will explore the operate phase and its unique context.

Operate

The operate phase is where design and engineering efforts are truly tested. During this operational period, security is a continuous process. It's not enough to deploy an SIS with state-of-the-art protection – if the system isn't maintained, monitored, and continuously improved, vulnerabilities can and will manifest. This necessitates a vigilant monitoring system capable of detecting, logging, and reacting to anomalous activities that could signal a breach. Regular system updates and patches are a must, with each deployment conducted under strict change management protocols.

Operation and maintenance

This stage of the ICS lifecycle involves the regular use and upkeep of the system. The focus is on maintaining continuous, safe, and efficient operation of the system according to its intended functionality. It involves routine inspections, periodic maintenance, troubleshooting, and repairs. It may also involve system monitoring, data collection, and reporting. These activities help to prevent unforeseen system downtime, improve system longevity, and maintain system performance at optimal levels.

Maintenance can be preventive, done routinely regardless of system performance, or reactive, also known as breakdown maintenance, performed in response to an identified issue. The goal is to reduce the risk of failure, machinery downtime, and productivity loss. Maintenance activities could range from something as simple as regular cleaning to more complex tasks such as replacing worn-out components.

Modification and retrofit

Over the lifecycle of an ICS, there may arise a need to modify or upgrade the system. This could be due to several reasons, such as changes in production requirements, advancements in technology, regulatory changes, or to mitigate potential risks that have been identified.

Some modifications may be minor, such as updating operating system versions or installing software patches. More extensive upgrades or retrofits might involve replacing obsolete hardware, upgrading or integrating systems for enhanced performance, or even comprehensive technology refreshes.

Modification and retrofit are crucial to keep the ICS current and competitive and to ensure its ongoing safety, efficiency, and compliance. Every modification requires a systematic process of planning, design, testing, and implementation to ensure it does not disrupt current operations or introduce new risks. After completion, it's beneficial to perform a post-implementation review to assess the effectiveness of the modification, draw lessons learned, and improve future projects.

Decommissioning

The decommissioning phase is the final stage of the lifecycle. It involves planning and carrying out the safe dismantling and removal of system components when they are no longer needed or can no longer be adequately maintained. This stage arises when a system has reached the end of its useful life, it's obsolete, or is being replaced by an upgraded system. Here are key activities in the decommissioning phase:

- This is the process of taking an ICS out of active service. It's a planned process and should be undertaken methodically to ensure a smooth transition. This process typically involves shutting down the system in an ordered manner to prevent any unexpected incidents or damages.

- It also includes disconnecting the system from any network interfaces, removing any stored information, de-energizing equipment, and deactivating software. Considerations need to be made for the impact on other systems and operations.

Depending on the nature of the ICS, decommissioning plans can also include preparations for the immediate replacement or upgrade of the system. This might involve data migration or other prep work to minimize downtime.

- The disposal process of an ICS should be managed carefully due to potential environmental and security concerns. It involves safely getting rid of physical hardware components, which may include servers, communication devices, controller modules, sensors, and so on. Depending on the environmental regulations, components may need to be disposed of in a specific manner to prevent environmental harm. Some parts may be able to be recycled or reused. For the disposal of storage devices, precautions must be taken to securely erase (that is, wiping and degaussing techniques) any sensitive data to prevent unauthorized access and to maintain data privacy. The disposal process may require thorough documentation for regulatory compliance.

The decommissioning phase, although at the end of the ICS lifecycle, is a crucial stage that requires careful planning and execution to prevent operational disruptions, environmental harm, and potential security risks.

In the following section, we will explore technology and product selection as crucial instruments for ensuring sustainable security and assurance.

Technology and product selection

The selection of security technologies for ICS production or safety critical-mission environment is critical. Organizations, ICS security professionals, and suppliers should be confident about product capabilities and the potential negative impact that they can have on live systems.

In recent years, there has been a growing recognition of the importance of security testing and evaluation for ICSs and SISs, not just as a requirement for vendors but also as a necessary practice for end users. Many organizations have implemented internal initiatives to regularly assess the security of their systems and ensure they meet industry standards and best practices. As a result, the field of ICS and SIS security testing and evaluation has expanded and become more important in the overall security landscape.

However, despite this progress, there are still challenges and areas for improvement in the field. One of the biggest challenges is the lack of understanding and standardization surrounding SIS security. Unlike other areas of cybersecurity, such as network or application security, where there are established frameworks and methodologies, SIS security is still a relatively new and evolving field. This means that there is often a lack of clear guidelines and best practices for conducting security testing and evaluation on these systems.

Testing ICSs is a multifaceted process, with each type of test targeting different aspects of system integrity and security.

The following list summarizes the various types of tests performed depending on the organization's risk appetite and asset criticality:

- **Penetration testing**: This is a simulated cyber attack against the ICS to check for exploitable vulnerabilities. Pen testing provides insight into the security posture of the system by attempting to breach various system parts using known vulnerabilities.

- **Vulnerability assessment**: Unlike penetration testing, vulnerability assessment does not exploit vulnerabilities but identifies, quantifies, and prioritizes potential vulnerabilities in the system. It's usually performed using automated scanning tools and presents a broader view of potential security flaws.

- **Source code review**: This is an examination of the application's source code to identify coding errors that could lead to security breaches. It's a thorough process that complements automated tools with manual review by an expert to ensure the code conforms to best practices and doesn't contain security defects.

- **Threat modeling**: This is a proactive approach that involves identifying and understanding potential threats and designing countermeasures to prevent or mitigate attacks. It involves reviewing the system architecture to anticipate where and how attackers might exploit it.

- **Reverse engineering**: This testing type is about disassembling the software to understand its components and their interrelationships. It can reveal insights into the system's functionality and potential weaknesses that might not be apparent through other testing methods. Reverse engineering may also be used to analyze malware that has infected the ICS.

These various types of tests are essential to ensuring the robustness, reliability, and security of ICSs. They help uncover potential issues early before commissioning the system and continue ensuring the security of these critical systems throughout their lifecycle.

We have covered the ICS lifecycle with a focus on secure by design, including asset and project lifecycle as it pertains to ICS. In the next section of this chapter, we will explore the need for CSMSs to organize, deploy, and improve security posture.

CSMSs

CSMSs serve as the strategic framework for safeguarding critical control systems from cyber threats. By implementing a CSMS, organizations systematize their approach to cybersecurity, bringing order and structure to their initiatives. A CSMS outlines policies, procedures, and controls that are essential for assessing risks, managing vulnerabilities, and responding to incidents. This organized approach is critical for prioritization – helping companies focus on the most crucial vulnerabilities and allocate resources effectively to areas with the highest potential impact. *Figure 5.2* shows the various components of a CSMS framework:

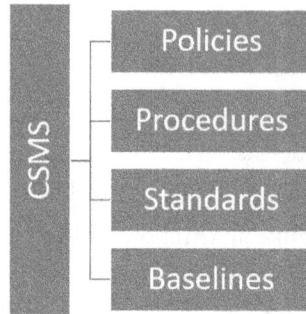

Figure 5.2 – CSMS framework

In the context of a CSMS framework, policies, procedures, standards, and baselines are crucial components that provide structure and governance to an organization's cybersecurity efforts:

- **Policies** are high-level documents that outline an organization's cybersecurity philosophy, objectives, and overarching commitments. They set the strategic direction and define the organization's stance on various cyber risks, creating a framework for decision-making and actions related to cybersecurity. Policies typically state what needs to be done but not how to do it, offering overall guidance to all levels of the organization.

- **Procedures** are step-by-step instructions that explain how to implement policies in operational terms. They provide detailed guidance to employees on specific actions they must take to comply with the cybersecurity policies. Often, procedures will cover aspects such as **Incident Response (IR)**, access control, and system maintenance, offering a clear path for routine and critical tasks.

- **Standards** are documented requirements that establish minimum acceptable limits, or benchmarks, for systems, processes, and controls within the organization. They help ensure consistency and reliability by defining specific technologies, configurations, and methodologies to be used. Standards often refer to industry-recognized best practices and are typically more technical and detailed than policies.

- **Baselines** set minimum security controls required for a particular system or class of systems. Baselines provide a foundation from which to evaluate and measure the security level of IT systems against a known set of criteria. They usually represent a risk-informed compromise between security needs and operational requirements, aiming to establish a known and manageable security posture across all systems within the organization.

Collectively, these elements work together to create a CSMS framework that helps organizations organize, standardize, and improve their security initiatives consistently and effectively while aligning them with business goals and regulatory requirements.

By having a CSMS in place, businesses can continuously monitor their ICS environments for emerging threats and apply a tactical mix of preventative and detective measures. Moreover, a CSMS fosters an environment of continuous improvement through regular reviews and updates to security practices, ensuring that the security posture evolves in tandem with the changing threat landscape. It also promotes better alignment between IT and OT units, creating a unified defense strategy that spans different segments of an organization. Ultimately, a CSMS enhances the resilience of ICSs by ensuring that security efforts are not sporadic or ad hoc but instead are driven by a comprehensive, well-organized strategy that aligns with the organization's overall risk management framework and business objectives.

A core aspect of an effective CSMS is developing a robust set of cybersecurity policies and procedures. These policies should align with industry standards and regulations, such as *IEC 62443* and *NIST SP 800-82*, and be specific to the unique requirements of the organization's ICS environment. Administrative controls, such as security awareness training, access management, and **IR Plans (IRPs)**, are essential tools to ensure adherence to these policies and procedures.

Furthermore, a well-defined CSMS governance structure is crucial to overseeing the implementation and enforcement of these policies and procedures across the organization. The implementation of strong technical countermeasures is also critical in protecting SISs and, by extension, the ICS environment. These measures include the deployment of firewalls, **Intrusion Detection and Prevention Systems (IDPSs)**, endpoint security, and network segmentation.

Firewalls act as the first line of defense by filtering and monitoring traffic entering and exiting the network. IDSs and **Intrusion Prevention Systems (IPSs)** can detect and prevent unauthorized access, malicious attacks, and suspicious behavior within the network. Endpoint security, in the form of antivirus software, can protect individual devices from malicious software and prevent malicious code from infecting the entire network. Network segmentation, where specific zones of the network are separated and isolated from each other, ensures that even if one section of the network is compromised, the entire system is not affected.

The *IEC 62443-2-1* standard specifies the requirements for establishing and maintaining an effective CSMS for **Industrial Automation and Control Systems (IACSs)**.

The following is an overview presented as a simplified process table:

Phase	Step	Description
1. Define Policies and Scope	a. Security Policy	Create a high-level organizational security policy that reflects cybersecurity objectives and commitments for IACS environments.
	b. Risk Assessment Policy	Develop policies for conducting and responding to risk assessments.
	c. Define CSMS Scope	Outline the assets, organizations, processes, and technologies that will be covered by the CSMS.

Phase	Step	Description
2. Assess Risks	a. Risk Identification	Identify cybersecurity risks to the IACS by understanding potential threats, vulnerabilities, and impacts.
	b. Risk Evaluation and Treatment	Evaluate the identified risks and formulate strategies to treat them according to the organization's risk tolerance.
3. Implement CSMS Functions	a. CSMS Control Types	Establish administrative, technical, and physical controls to mitigate identified risks.
	b. Security Levels	Define and implement security levels based on the severity of risks and the criticality of assets.
	c. System Protection	Apply protective measures to IACS components – segmentation, application whitelisting, secure remote access protocols, and so on.
4. Establish CSMS Processes	a. Incident Response	Develop procedures for detecting, responding to, and recovering from security incidents.
	b. Business Continuity	Ensure that backup and recovery plans are in place to maintain or quickly resume business operations.
	c. Training and Awareness	Implement ongoing training and awareness programs for individuals involved in the operation and security of IACSs.
5. Monitor and Improve	a. Performance Measurement	Establish indicators to measure the effectiveness and performance of the CSMS.
	b. Audits and Reviews	Perform regular audits and reviews of the CSMS to ensure it is functioning as intended and to uncover any areas for improvement.
	c. Continuous Improvement	Use the insights from performance monitoring and audits to refine and enhance CSMS processes and controls.
6. Maintain Compliance	a. Documentation and Records	Keep detailed documentation and records to support CSMS process integrity, including policy enforcement, training, and incident handling.
	b. Legal and Regulatory Requirements	Stay abreast of and ensure compliance with all applicable cybersecurity laws, regulations, and industry standards.
	c. Certification and Assessment Bodies	Work with external certification and assessment bodies, as needed, to verify compliance with security standards such as IEC 62443-2-1.

Table 5.1 – IEC 62443 CSMS

This table simplifies the complex processes and functions described in *IEC 62443-2-1* by condensing them into a series of steps that contribute to establishing a comprehensive CSMS. It allows organizations to visualize the journey from policy formulation to continuous improvement and compliance maintenance. Each phase builds on the previous to create a dynamic, responsive CSMS that not only protects vital ICS assets but also evolves with the changing cybersecurity landscape.

We explained the foundation with CSMSs; let's now explore how integral security brings all the elements together to provide solid security.

SIS – The need for integral security

An SIS requires integral security that encompasses a combination of physical, technical, managerial, and mechanical countermeasures to protect against cyber threats. Regular risk assessments, robust cybersecurity policies and procedures, and the deployment of technical, physical, and mechanical controls are all vital elements that must be considered to achieve a high level of resilience and ensure the safe and secure operation of ICSs.

In order to ensure comprehensive security for SISs, a meticulous and expansive approach is required, weaving together managerial, physical, digital, and mechanical threads into a fortified tapestry of protection. This multidimensional defense initiative begins with a foundation of solid managerial controls – a suite of policies and procedures recognized as the lifeblood of SIS security governance.

Moreover, these authoritative documents must articulate a clear vision for security objectives, assign explicit responsibilities to stakeholders, and delineate step-by-step processes for managing risks throughout the SIS lifecycle. Policies and procedures act as the guiding framework for conducting regular risk assessments, which are vital to identifying potential vulnerabilities and implementing the appropriate safeguards across all facets of SIS infrastructure.

Furthermore, bolstering this managerial backbone are robust physical security measures, designed to thwart unauthorized access and mitigate threats at the most tangible level. High-security locks, reinforced access points, physically segregated network interfaces, and guarded facility perimeters erect a formidable barrier against intrusion. Surveillance systems, coupled with motion-detection lighting and alarm systems, act as both deterrent and sentinel, keeping watchful eyes on SIS components that operate in critical, often hazardous environments. Physical security also extends to environmental considerations, where controls ensure stability against natural or man-made disasters, shielding SISs from extremes in temperature, moisture, or vibrations that could compromise their operational integrity.

These physical defenses are complemented by a digital fortress of cybersecurity measures – a multifaceted shield against the spectral threats of the virtual domain. Network security architectures are constructed with a **Defense-in-Depth (DiD)** mindset, employing redundant layers of firewalls, **Demilitarized Zones (DMZs)**, and stringent network segmentation to confine and control data flows. IDPSs act in real time to fend off attempted breaches, while advanced cybersecurity solutions such as whitelisting and behavioral analytics guard against sophisticated malware and zero-day exploits. Ethical hacking initiatives and red team exercises refine digital defenses continuously, while encryption protocols keep sensitive SIS data obscured from prying eyes, whether in transit or at rest.

Parallel to these initiatives, mechanical security controls exist as a resolute physical counterbalance to digital measures. These controls encompass safeguards such as fail-safes, mechanical locks, tamper-evident seals, and exhaustive **Emergency Shutdown (ESD)** procedures – each designed to ensure SIS functionality is preserved, even in the event of irregularities or compromises elsewhere in the system. Mechanical integrity programs emphasize regular maintenance, validation, and rigorous testing of SIS components to forestall physical degradation or malfunctions that might otherwise go unnoticed until catastrophic failure occurs.

Finally, this holistic security approach champions collaboration between cross-disciplinary teams, binding engineering expertise with IT acumen to forge a resilient, integrated defense strategy. Routine drills, simulations, and cybersecurity awareness training instill a culture of security mindfulness, making each employee an active participant in the defense of the SIS.

Managerial policies enable rapid IR, ensuring that when anomalies are detected, swift, protocol-driven action minimizes harm and facilitates recovery. By intertwining managerial controls with physical, digital, and mechanical safeguards, organizations ensure that SISs are not only compliant with industry standards but are also resilient against emergent and ever-evolving threats – securing the continuity of industrial processes and the safety of human and environmental assets that rely on them. *Figure 5.3* represents the various elements of SIS integral security:

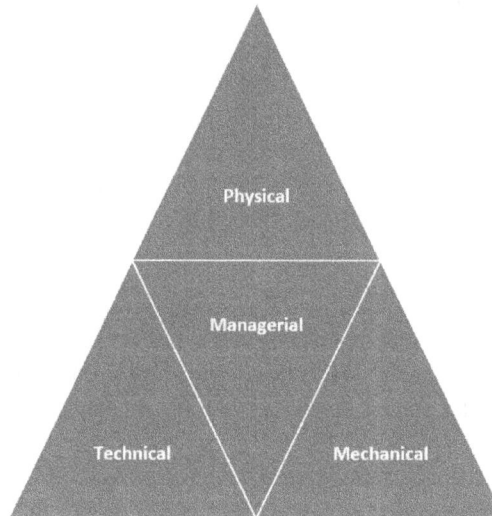

Figure 5.3 – SIS integral security

Let's examine the elements of physical security and its fundamental role as part of integral security.

Physical

Physical security controls to protect SIS assets are designed to prevent unauthorized access to crucial system components and to safeguard against environmental hazards and potential physical tampering. These controls are integral components of a comprehensive security strategy. Here are several examples of physical security controls that can be implemented to protect SIS assets:

- **Access controls**: Implementing card readers, biometrics, PIN codes, or other security devices at entry points to restrict access to facilities housing SIS components to authorized personnel only

- **Security fencing**: Erecting fences around critical infrastructure to deter unauthorized entry and encroachment, enhancing the protection of external SIS components

- **Surveillance cameras**: Installing **closed-circuit television** (**CCTV**) and surveillance cameras to monitor facilities and deter tampering or vandalism, as well as to provide evidence in the event of a security incident

- **Security guards**: Employing on-site security personnel to conduct patrols, oversee entry points, and respond to security alerts provides a direct line of defense against unauthorized access

- **Lighting**: Ensuring adequate lighting around key SIS-related facilities can deter intruders and improve visibility for monitoring systems, reducing the risk of unauthorized activities during non-operational hours

- **Locked enclosures**: Using locked cabinets, cages, or enclosures to physically secure SIS hardware and field devices from tampering, unauthorized adjustments, or theft

- **Tamper-evident seals**: Applying tamper-evident seals on critical equipment so that any unauthorized access can be quickly identified

- **Environmental controls**: Implementing environmental monitoring and control systems to protect SIS assets from extreme temperatures, humidity, dust, and corrosive environments that could affect performance or cause damage

- **Fire suppression systems**: Installing appropriate fire detection and suppression systems to protect against potential fire hazards that might critically affect SIS operation

- **Equipment redundancy**: Utilizing redundant systems and physical separation of critical components to ensure that failure or compromise of one does not impact the availability and functionality of the SIS

- **Signage**: Using clear and prominent signage to identify restricted areas and deter trespassing, as well as to guide authorized personnel in maintaining safety protocols

- **Vehicle barriers**: Employing vehicle barriers such as bollards, gates, or tire spikes at facility entry points to prevent unauthorized vehicle access or mitigate vehicular threats

- **Mantraps and airlocks**: For highly sensitive areas, using a mantrap or airlock can ensure that two sets of doors cannot be open at the same time, controlling individual access to SIS assets

By integrating these physical security controls into a broader security management plan, organizations can significantly mitigate risks associated with unauthorized physical access, environmental factors, and intentional tampering. These measures contribute to the overall resilience and reliability of SIS operations, which are crucial for ensuring the safety of industrial processes.

Technical

ICS technical security controls are critical measures and practices designed to safeguard crucial infrastructure against a multitude of cyber threats. Each component is integral to maintaining the integrity, confidentiality, and availability of ICS environments:

- **Asset management**: This entails a thorough inventory and management of all ICS components, providing visibility into the assets owned, their location, and their relevance to the operational process. It forms the bedrock of ICS security by ensuring all assets are accounted for and assessed for vulnerabilities.

- **Access control**: Implementing strict controls that govern who can access the ICS environment, detailing their level of interaction with the system, and tracking their activities to prevent unauthorized use and potential system breaches.

- **System hardening**: The process of securing systems by reducing their attack surface, which includes patching software, disabling unnecessary services, removing redundant user accounts, and implementing least-privilege principles.

- **Network security and architecture**: Establishing a secure network design that isolates the control systems environment from other networks, deploying firewalls, DMZs, and IDPSs to thwart unauthorized access and network-based attacks.

- **Backup and recovery**: Critical for resilience in the face of adverse events, this involves creating accurate and retrievable copies of system data and configurations to ensure continuity of operations, even after a malicious cyber incident.

- **Logging and monitoring**: Continuous oversight of ICS activities through the collection, analysis, and archiving of logs to detect and escalate abnormal events or potential security incidents rapidly.

- **Vulnerability management**: Identifying, assessing, and mitigating vulnerabilities within the ICS by regularly updating and patching systems, conducting vulnerability assessments, and employing proactive threat-hunting initiatives.

- **Endpoint protection**: Protecting the points of access to ICS networks – such as workstations, servers, and mobile devices – by employing antivirus software, application whitelisting, and device control measures to prevent malware infection and ensure only trusted software runs.

- **Remote access**: Safeguarding all external access points to the ICS network by using secure methods, such as **virtual private networks** (**VPNs**) with strong encryption, **two-factor authentication** (**2FA**), and comprehensive access controls to avert unauthorized remote entry.

- **IR and forensics**: Defining emergency response procedures to promptly address security breaches, manage crisis situations, conduct digital forensic analysis to uncover attack vectors, and gather evidence for post-incident analysis.

- **Cybersecurity documentation**: Maintaining a repository of well-structured and detailed cybersecurity governance and operational documents supports consistent implementation of security protocols, personnel training, and compliance with regulatory standards.

These ICS technical security controls, when effectively implemented, ensure a robust security framework capable of withstanding and evolving in the face of persistent cyber risks, thus preserving the integrity and operational stability of critical infrastructure systems.

Blind spots

In the domain of SIS cybersecurity, multiple blind spots can exist that may lead to vulnerabilities if overlooked. Identifying and addressing these potential oversights is crucial to ensure the robustness of SIS cybersecurity measures. Here is a list of possible blind spots:

- **Legacy systems**: Older SISs that may not have been designed with cybersecurity in mind, lacking the ability to integrate modern security features

- **Insufficient segregation**: Inadequate network segmentation between IT and OT networks, allowing potential pathways for attackers

- **Vendor-specific vulnerabilities**: Proprietary systems that may have undocumented vulnerabilities or reliance on a single vendor for patches and updates

- **Lack of visibility**: Inadequate monitoring and detection capabilities that fail to provide visibility into SIS network traffic and potential malicious activities

- **Insufficient access controls**: Weak authentication mechanisms and overly permissive access that allow unauthorized changes to the SIS

- **Unpatched software**: Delayed or unapplied software and firmware updates that leave known vulnerabilities open to exploitation

- **Physical security overlooked**: Neglecting physical security aspects of SIS components, which could be exploited to gain unauthorized access

- **Supply chain risks**: Vulnerabilities introduced through third-party suppliers, including software, hardware, and service providers

- **Personnel factor**: Lack of employee training and awareness, leading to potential inadvertent insider threats or inadequate response to incidents

- **Inadequate change management**: Poor change management practices leading to unauthorized or unrecorded modifications that can affect SIS operations

- **Integration of Internet of Things (IoT) devices**: Increased use of IoT devices without proper security consideration can introduce new vulnerabilities

- **Cyber-physical interaction risks**: Ignoring the interplay between cybersecurity threats and physical safety risks that can have real-life implications

- **Insufficient IRP**: An outdated or untested IRP can result in disorganized handling of a cyber incident

- **Mobile and remote access risks**: Weak controls over mobile device management and remote access can expose SISs to additional threats

- **Compliance-only mindset**: Relying solely on compliance with standards without implementing a proactive and adaptive security posture

- **Cybersecurity skills gap**: Shortage of personnel with both cybersecurity- and SIS-specific knowledge, leading to potential oversights

- **Data integrity oversight**: Inadequate protection mechanisms to ensure the integrity of the data used by the SIS for making safety-related decisions

- **Insecure communication protocols**: Use of non-encrypted or weakly encrypted communication protocols for SIS data transfer

- **Lack of forensic capability**: Not having the necessary tools or procedures in place for post-incident forensics and analysis to learn from events and strengthen defenses

- **Overreliance on third-party security**: Assuming that third-party services, such as cloud providers, are solely responsible for the security of services they host

By accounting for these potential blind spots, organizations can take proactive steps toward the development of a comprehensive SIS cybersecurity strategy that mitigates risks and enhances overall system resilience.

Mechanical

Mechanical controls in SISs serve as essential safeguards by providing physical barriers, fail-safes, or intervention mechanisms that enhance operational security and safety. Here are several examples of mechanical controls that can safeguard SIS assets:

- **Safety valves**: These are critical for controlling the flow of materials (for example, gas, fluids) in industrial processes. Safety valves are designed to open automatically to relieve pressure and prevent a potential overpressure event that could lead to equipment failure or an explosion.

- **Pressure relief devices**: Similar to safety valves, pressure relief devices act as a mechanical safeguard to protect systems from overpressure scenarios by venting excess pressure to a safe location.

- **ESD valves**: ESD valves can rapidly isolate sections of a process or shut down the operation completely in response to a hazardous event, preventing escalation and protecting both personnel and the environment.

- **Transmitters with mechanical gauges**: While transmitters provide electrical signals corresponding to various process conditions, mechanical gauges offer a local, direct measurement that can be verified visually, serving as a backup in case of electronic failure.

- **Key switches**: Key switches provide a physical means to enable or disable certain SIS functions. Only individuals with the appropriate key can alter the status of a system or override an automatic function, ensuring strict access control.

- **Lock-Out/Tag-Out (LOTO) systems**: LOTO systems physically lock control devices or power sources during maintenance or servicing to prevent accidental startup or release of hazardous energy.

- **Manual overrides**: Manually operated switches or levers that allow operators to take direct control of a system in case of an emergency or when automatic systems fail.

- **Actuators**: Mechanical actuators on valves and other components enable automatic or manual operation to control the flow of materials and ensure safe process conditions.

- **Bursting discs**: These are pressure relief devices that rupture at a pre-determined pressure or temperature to provide a safety release mechanism in the case of overpressure.

- **Flow switches**: These devices mechanically respond to changes in flow rate, triggering alarms or activating shutdown protocols if flow is detected outside of safe operating parameters.

- **Tamper-proof seals**: Seals that indicate if equipment or access points have been tampered with or accessed without permission.

- **Position switches**: Mechanical devices that confirm the presence or position of a component (such as a valve) and can be used to ensure it is in the correct state before process operations proceed.

- **Mechanical interlocks**: These are devices that prevent certain actions from being taken unless a pre-set condition is achieved – such as certain valves being open or closed – ensuring that operations proceed in a safe, sequential manner.

- **Explosion-proof enclosures**: Physical enclosures designed to contain and prevent the propagation of an explosion caused by sparks or electrical failures within SIS components.

By incorporating these mechanical controls, organizations can create multiple layers of protection to respond to potential hazards, enhancing the safety and reliability of SISs and the processes they control.

Digital versus manual controls

The cyber attack on Ukraine's power plant back in 2015 serves as a somber reminder of the vulnerabilities inherent in relying solely on digital controls for critical infrastructure. The assault, which caused blackouts and significant service interruptions, highlighted the strategic value of such infrastructure as a target for adversaries seeking to disrupt a nation's essential functions.

One of the key lessons learned from this incident is the importance of not depending exclusively on digital control systems. While these systems offer convenience and efficiency, they can also be susceptible to cyber threats such as hacking, malware, and ransomware. Consequently, there is a need for redundancy through mechanical and manual controls that can serve as reliable backups in case digital systems fail or are compromised. Such controls can provide a fail-safe mechanism to shut down operations or switch to a manual mode, ensuring that operators maintain control over critical processes even when cyber defenses are breached.

The attack on Ukraine also reinforces the necessity for robust contingency plans that include manual overrides and physical security measures. These plans should be ingrained in the regular training of staff so that immediate and effective manual intervention can be made during cyber incidents. Moreover, it emphasizes the need for a holistic security approach that incorporates both cyber and physical protections to prevent unauthorized access of any kind.

To sum it up, the Ukraine power plant attack not only demonstrated the disruptive potential of cyber warfare but also served as a catalyst for reassessing how critical infrastructure is safeguarded. It urged industries and governments worldwide to recognize the indispensable role of layered defenses, where digital, mechanical, and manual controls work in concert to ensure continuity, resilience, and safety of essential services.

Managerial

Managerial controls for SISs consist of a structured set of policies, procedures, and processes that govern the operational integrity, security, and maintenance of these critical systems. These controls are established to ensure that all safety-related activities are performed consistently and effectively, minimizing risk and enhancing the SIS's protective functions. Next are some examples of managerial controls designed to protect SIS assets:

- **Security policy**: Comprehensive security policies specific to SIS assets, outlining the organization's stance on maintaining the confidentiality, integrity, and availability of SIS information and resources

- **Change management procedures**: Rigorous procedures to manage changes to SIS configurations, hardware, and software, ensuring that all changes are assessed for risks, appropriately tested, documented, and approved before implementation

- **Access control policy**: Policies that define who has access to SIS components, under what conditions they can have such access, and what level of access is to be granted to individual users, including contractors and vendors

- **Maintenance and testing procedures**: Periodic maintenance and thorough testing procedures for SIS equipment to ensure ongoing reliability, including calibration of sensors, testing of fail-safes, and verification of alarm systems

- **IRP**: A well-defined IRP that sets forth the process for responding to SIS-related security incidents, containing instructions for escalation, communication, investigation, and post-incident review

- **Training and awareness programs**: Regular training programs for operators, maintenance personnel, and other staff members involved with SISs to reinforce awareness of safety policies, proper procedures, and the importance of security measures

- **Audit and review processes**: Regular audits of SIS security practices and an ongoing review process to ensure that policies and procedures remain effective, are adhered to, and are updated in response to new threats or changes in the operating environment

- **Risk management framework**: A comprehensive risk management framework for analyzing and prioritizing risks to SIS assets, along with strategies for risk mitigation, transfer, acceptance, or avoidance

- **Business Continuity Plans And Disaster Recovery Plans (BCPs and DRPs)**: Contingency planning for SIS assets to ensure that operations can continue or be quickly restored following disruptive events, ensuring minimal impact on process safety

- **Vendor and contractor management**: Policies for managing third-party access to SISs, including contractor selection criteria, security requirements for external vendors, and supervision during on-site activities

- **Performance measurement metrics**: Meticulously crafted **Key Performance Indicators (KPIs)** for the SIS security posture that serve to evaluate compliance with policies and effectiveness in managing risks

- **User authentication and authorization protocols**: Managerial controls over authentication and authorization mechanisms, ensuring that **Multi-Factor Authentication (MFA)**, role-based access, and least-privilege concepts are consistently applied

By implementing these managerial controls, an organization sets forth a controlled environment where SIS assets are used and maintained in a manner that prioritizes safety and security. It creates a culture of accountability and vigilance that is critical in mitigating risks and safeguarding the integrity and functionality of SISs.

Emerging compliance and management accountability

Management accountability in the context of protecting critical infrastructure is a vital aspect of corporate governance. When leadership fails to implement good practices for securing assets, it can have grave consequences, not just for the organization but for national security, economic stability, and public welfare. The obligation of management to safeguard their assets extends beyond moral responsibility—it is also enshrined in law through various regulations and legislation.

In the European Union, the **Network and Information Systems (NIS2)** Directive represents a key legislative framework that obligates operators of essential services and **Digital Service Providers (DSPs)** to take appropriate security measures and report serious cyber incidents. This directive holds management accountable by requiring them to instill strong cybersecurity practices.

Similarly, in the United States, legislation such as the forthcoming **Cyber Resilience Act** is expected to set standards and obligations for entities in critical sectors. While the legislation is still evolving, the principle is that it would impose strict requirements for risk management, regular assessments, and mitigative actions against cybersecurity threats, with possible penalties for non-compliance.

In addition, a **Software Bill of Materials** (**SBOM**) is one of the most important initiatives launched to mitigate increasing supply chain issues and product security.

These regulatory frameworks underscore the importance of management's role in ensuring that robust cybersecurity defenses are in place. It sends a clear message that senior executives and boards can no longer be passive about cyber threats but must actively drive the cybersecurity agenda, enforcing policies and practices that protect critical infrastructure from evolving risks. Failure to do so can lead to reputational damage, legal repercussions, and significant financial losses, not to mention potential harm to national interests and public safety.

Besides the managerial controls that have been discussed, there are operational policies and procedures that are very specific to the ICS space and in particular for safety critical systems. They are somehow different from typical policies and procedures deployed in IT environments.

We will explain the specification of operational policies and procedures in the next section.

Operational policies and procedures

Operational security policies, procedures, and standards are commonly utilized in IT environments. However, in the realm of ICSs, this area is relatively new and is evolving rapidly due to various factors. In the past, functional safety and physical security were the main areas of focus for SIS practices. However, there have been instances where efforts from vendors and asset owners have aimed to increase maturity and professionalism in this field. ICSs and SISs are integral components of modern-day industrial operations, playing a crucial role in maintaining efficient and safe processes.

In recent years, with the increasing digitization and connectivity of these systems, the issue of cybersecurity has become a significant concern. It is imperative for companies to implement strict operational policies and procedures to ensure the protection of these critical systems from cyber threats. Let us explore in detail the various aspects of ICS and SIS cybersecurity operational policies and procedures:

- **Permit to work (PTW)**: A PTW is a standard procedure that regulates and controls potentially hazardous activities within a facility. It is a crucial component of ICS and SIS cybersecurity operational policies as it ensures that proper authorization and supervision are in place for any changes or maintenance work being carried out on these systems. Companies must implement a strict PTW process, including thorough risk assessments and training for personnel involved in any work on the systems.

- **Management of change (MOC)**: With the rapid advancements in technology, it is necessary to have a robust system in place to manage any changes made to the ICS and SIS. This includes software updates, hardware replacement, and expansions. A well-defined MOC process should be in place to ensure that any alterations to the systems are carried out with minimal risk and maximum efficiency. This process should involve a thorough review of the impact of changes on the cybersecurity of the ICS and SIS.

- **Asset integrity and reliability**: Maintaining the integrity and reliability of assets is crucial in ensuring the smooth operation of ICSs and SISs. Cyber threats can compromise the functionality of these systems, leading to potential safety hazards and operational disruptions. To mitigate such risks, operational policies and procedures should include regular monitoring, maintenance, and testing of assets to ensure their proper functioning and identify any potential vulnerabilities. This process can also include implementing IDPSs to enhance the security of assets.

- **Workforce development**: Human error is a significant contributing factor to cybersecurity breaches in ICSs and SISs. Therefore, it is essential to have a well-trained and knowledgeable workforce who are aware of potential risks and are equipped to handle them. Companies should invest in continuous training programs for their employees and have a clearly defined hiring process that includes background checks and security screenings. This will ensure a competent and secure workforce that can effectively handle cybersecurity threats.

- **Remote access**: Remote access to ICSs and SISs can offer many benefits, such as increased efficiency and flexibility. However, it also poses a significant cybersecurity risk. Operational policies and procedures should strictly regulate remote access to these systems, including strong authentication measures, regular audits, and proper training for remote personnel. It may also be necessary to implement VPNs and other encryption methods to secure remote connections.

- **Operational readiness**: Ensuring that ICSs and SISs are continuously operational is crucial. Any disruption in these systems can lead to significant safety and financial hazards. Therefore, it is essential to have operational readiness policies and procedures in place to ensure the timely diagnosis and resolution of any issues. This can include continuous monitoring, backup systems, and DRPs.

- **Maintenance**: Regular and timely maintenance of ICSs and SISs is vital in identifying and addressing any potential cyber threats. Companies should have a detailed maintenance schedule in place that covers all aspects of these systems, such as software updates, hardware maintenance, and cybersecurity checks. This process should also include thorough risk assessments during maintenance to identify any potential security vulnerabilities.

- **Inspection and audit**: Inspection and audit procedures must be an essential part of any ICS and SIS cybersecurity operational policies. These processes should include regular checks and evaluations of all security measures, as well as emergency drills to test the effectiveness of cybersecurity protocols in place. Any identified gaps or weaknesses should be promptly addressed and fixed to enhance the overall security posture.

- **Obsolescence**: As technology evolves, it is crucial to keep up with the latest updates and advancements to ensure the security of ICSs and SISs. Companies should regularly review the hardware and software used in these systems and have a process in place to replace and upgrade obsolete components. This will not only enhance the cybersecurity of the systems but also ensure their longevity and reliability.

- **Bypasses**: In some cases, due to technical issues or emergencies, bypassing certain security measures may become necessary. However, this should be strictly regulated and documented through a bypass management process. It should involve risk assessments, approval from higher management, and proper monitoring to ensure that the bypass is only temporary and necessary.

- **Tools**: Having the right tools and technologies is crucial in maintaining the cybersecurity of ICSs and SISs. Companies should invest in the latest and most robust cybersecurity tools, such as firewalls, IDSs, and antivirus software, to safeguard their systems from potential threats. Furthermore, policies should be in place to regularly review and update these tools to match the evolving threat landscape.

In conclusion, implementing strict operational policies and procedures is essential in ensuring the cybersecurity of ICSs and SISs. It is a continuous process that requires regular reviews and updates to keep up with the rapidly evolving threat landscape. By following the aforementioned guidelines, companies can mitigate cyber risks and maintain the efficient and safe operation of their systems.

The next section will examine a very sensitive but critical element of successful security within the ICS environment.

GOM

A GOM is a crucial concept in the field of OT and ICSs. It refers to the framework for managing and overseeing processes, policies, and procedures that govern cross-functional relationships and responsibilities between IT and OT. A GOM is particularly important in the OT and ICS environment because these systems are responsible for controlling and monitoring mission critical industrial processes. In this section, we will explore the GOM and its key elements, with a focus on roles and responsibilities for IT and OT, as well as the segregation between control functions (that is, **Basic Process Control Systems (BPCSs)**, **Distributed Control Systems (DCSs)**, and safety functions (that is, SISs). We will also discuss challenges, opportunities, trends, and maturity levels associated with implementing a GOM in the OT and ICS environment. The following diagram depicts the demarcation point between IT and OT systems:

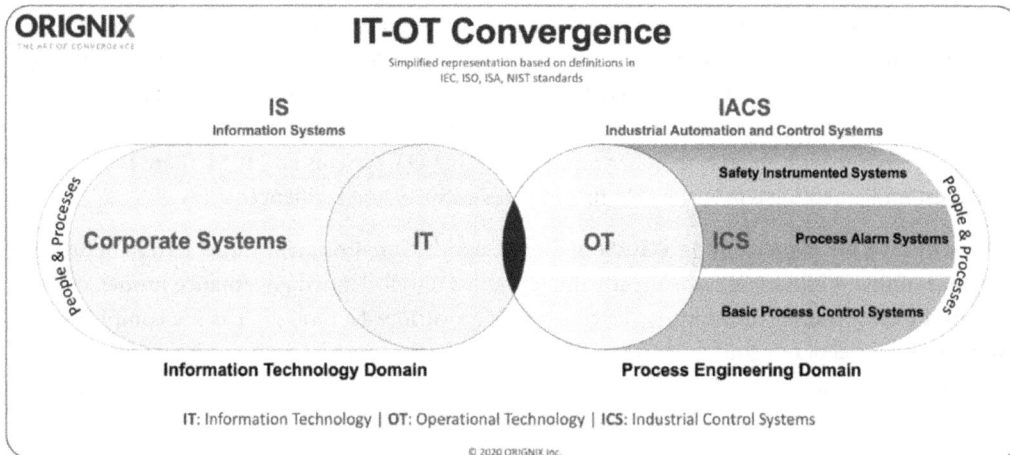

Figure 5.4 – IT/OT demarcation point (https://orignix.com/what-is-it-ot-convergence/)

A well-defined GOM in the OT and ICS environment is necessary to ensure effective management and control of these critical systems. One of the key elements of the GOM is roles and responsibilities. In this context, it implies clearly defining the roles and responsibilities of both IT and OT teams in managing and maintaining OT and ICSs. The GOM should outline specific tasks and decision-making authorities for each team to appropriately manage systems, mitigate risks, and respond to security incidents. Additionally, it is essential to establish a clear segregation between the control function (that is, the DCS) and safety functions (that is, the SIS) to maintain the integrity and reliability of the systems. The control function is responsible for monitoring and controlling industrial processes, while the safety functions are crucial for ensuring the safety and protection of operations. The GOM should clearly define roles and responsibilities for each function, including handoff points between them to ensure proper coordination and communication.

While implementing a GOM in the OT and ICS environment has its benefits, it brings along its set of challenges. One significant challenge is managing the complexity of the systems and processes involved. The OT and ICS environment typically comprises a wide range of interconnected and heterogeneous systems, making it difficult to define a one-size-fits-all approach. Additionally, the responsibility of managing these systems is often shared between IT and OT teams, leading to potential conflicts and misaligned priorities. Another significant challenge is the constantly evolving threat landscape, which requires continuous monitoring and updating of the GOM to address new risks and vulnerabilities.

However, a well-designed GOM also presents opportunities for improved efficiency, streamlined processes, and enhanced security. A defined governance model that clearly outlines roles, responsibilities, and handoff points can help in effective risk management, reducing downtime, and improving response to security incidents. It also allows for better coordination and communication between IT and OT teams, fostering collaboration and promoting a shared understanding of the systems' objectives. Additionally, a well-defined GOM can help organizations stay compliant with regulatory requirements, ensuring the safety and security of critical infrastructure.

The trend in the OT and ICS environment is toward greater integration and convergence of IT and OT systems. This trend is driven by the need to improve efficiency, reduce costs, and enhance operational capabilities. As a result, there is an increased focus on implementing a GOM that enables seamless collaboration between the two teams while maintaining segregation between control and safety functions. Moreover, with the rise in cyber threats targeting OT systems and ICSs, there is a greater emphasis on implementing a robust GOM that ensures security and resilience.

The maturity level for implementing a GOM in the OT and ICS environment varies across industries and organizations. While some have already implemented a well-defined governance model, others are still in the early stages. The maturity level depends on various factors, such as the complexity of the systems, the level of integration between IT and OT, and the organization's readiness to embrace change. To achieve a higher level of maturity, organizations need to invest in developing a comprehensive GOM that considers the unique requirements of OT systems and ICSs.

A governance model for an SIS within the context of ICS cybersecurity is a structured framework that defines roles, responsibilities, and processes to ensure the effective protection of these critical components against cyber threats. This model is crucial for establishing clear accountability and promoting a culture of security within an organization.

At the heart of this governance model is the alignment of SIS cybersecurity objectives with the organization's broader risk management and business goals. This involves senior management taking an active role in endorsing cybersecurity initiatives and providing the necessary resources to support them.

Key aspects of the governance model typically include the following:

- **Leadership and commitment**: Top executives are accountable for cybersecurity, demonstrating their commitment through policy establishment, consistent communication, and ensuring the integration of cybersecurity into business and operational strategies

- **Roles and responsibilities**: Clear definitions of cybersecurity roles across the organization, detailing who is responsible for implementing, managing, and maintaining SIS cybersecurity measures

- **Policy framework**: Establishment of comprehensive cybersecurity policies that are tailored to the SIS environment and support regulatory compliance efforts

- **Risk management**: An iterative risk management process that identifies, assesses, and treats cybersecurity risks to the SIS, aligning with the overall risk management strategy of the organization

- **Resource allocation**: Ensuring that appropriate resources – including tools, personnel, and training – are allocated for cybersecurity defenses and activities related to the SIS

- **Performance monitoring**: Regular assessment and review of the SIS cybersecurity posture to gauge the performance of implemented controls and the effectiveness of the governance model

- **IR and recovery**: A well-structured IRP that outlines procedures for managing and recovering from cybersecurity events that impact the SIS

- **Continuous improvement**: Embedding a culture of continual learning and adaptation within the governance model to evolve with emerging cybersecurity threats and technological advancements

- **Stakeholder communication**: Effective communication with internal and external stakeholders, including regulatory bodies, to keep them informed about SIS cybersecurity matters and initiatives

- **Documentation and record keeping**: Meticulous documentation of governance activities, cybersecurity policies, risk assessments, and incident handling for accountability, compliance, and audit trails

By subscribing to this governance model, organizations ensure that their SIS, which is crucial for controlling hazardous processes, is well protected from cyber intrusions that could lead to dangerous failures or accidents. This governance framework facilitates a resilient and secure SIS within the larger ICS cybersecurity landscape. *Figure 5.5* provides an overview of governance for IT and OT, as well as control and safety functions:

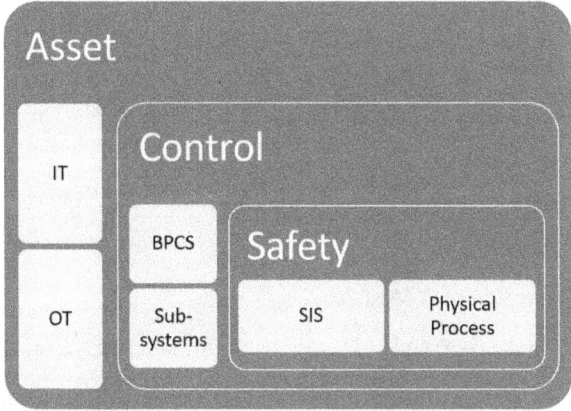

Figure 5.5 – Overview of IT/OT and control/safety demarcation points

In conclusion, the governance operating model is a critical framework for managing the relationship between IT and OT in the OT and ICS environment. It requires a clear definition of roles and responsibilities, proper segregation between control and safety functions, and continuous adaptation to the evolving threat landscape. Although there are challenges, a well-designed GOM presents opportunities for improved efficiency, streamlined processes, and enhanced security. The trend toward greater integration between IT and OT systems and the growing focus on security indicate the increasing importance of implementing a mature GOM in the OT and ICS environment.

Next is an example of SIS users that can be found in implementations:

Security Level	Allowed Access
View	This user has no rights, but the user can page through displays
SIS Operate	The operator has all View rights plus the following: • Alarm – Acknowledge • Request a Reset and Reset Permit
SIS Supervise	The supervisor has all operator rights plus the following: • Alarm – Settings, Suppress • Request a Bypass
SIS Engineer	The engineer has all supervisor rights plus the following: • Trip and Pre-Trip Limit changes • Trip Delay Time and Normal Delay Time Stable Time, Reminder Time, and so on • Alarm – Enable / Disable (SIS Restricted Control key) • Configure SIS • Download SIS Modules • Simulate – Enable / Disable (only available for Virtual logic solver)
End User Administrator	The end-user administrator has all engineer rights plus the following: • System maintenance • All other settings
Vendor Administrator	The vendor administrator has the same rights as the end-user administrator.

Table 5.2 – Example of security access level for an SIS

The next section highlights the importance of operation and cybersecurity maintenance.

Operation and cybersecurity maintenance

Cybersecurity maintenance for ICSs and SISs is an integral part of ensuring the safety and security of operations, and it requires a disciplined, continuous management approach. Regular maintenance activities should cover a broad spectrum of tasks aimed at preserving the integrity, availability, and confidentiality of system components and the data they process.

In this context, the three elements addressed next are essential for safe operation at all times.

Hazard

In order to ensure a predictable and reliable behavior of the ICS, it is important that all maintenance and modifications of ICSs are executed by authorized and qualified personnel in a structured and defined way as laid down in this procedure. Non-compliance with this procedure may lead to the following:

- Contradictory information in operating, engineering, and asset management environments
- Corruption of control systems databases
- Unreliable or failing hardware
- Failing systems communications

This could result in "misoperation" (production loss, trips), loss of redundancy, or even complete failure of monitoring, control, or safeguarding functions. In the worst case, this could have severe consequences for personnel, equipment, and/or environments.

Criticality

Criticality is a measure of urgency for the use of the equipment concerned. Criticality relates directly to the adverse effects that non-availability, either revealed or unrevealed, may have on the following:

- Plant and personal safety
- The environment
- The integrity of plant equipment
- Control of product inventory
- Plant operability
- Statutory and contractual obligations
- Well-being

Reliability

Reliability is the probability that the equipment concerned will function within its intended performance envelope for a specified period of time without failure under the expected environmental and operational stresses encountered in its service.

Here are aspects that should be taken into consideration when it comes to operation and cybersecurity maintenance:

- **Regular monitoring and inspection**: Routine inspections are integral to properly maintaining any ICS. Regular checks of hardware components, software functions, and network connections can help identify issues before they become serious problems. It's crucial to use appropriate monitoring tools that provide real-time visibility of system operations and generate alerts for out-of-norm conditions.

- **Preventive maintenance**: This involves scheduled, routine maintenance activities aimed at preventing system breakdowns and failures. It could include tasks such as cleaning, lubricating, adjusting, and making minor repairs. This proactive approach can reduce unexpected downtime and improve the system's lifespan.

- **Reactive maintenance**: Also known as breakdown maintenance, it's the repair work performed when a system component fails. It's crucial to quickly diagnose the problem, fix it, and get the system back to operation, minimizing disruption.

- **Training and skill development**: A knowledgeable and skilled workforce is vital to properly run and maintain an ICS. Regular training programs, skill development courses, and awareness initiatives can ensure the team is capable of operating the system efficiently, can identify problems early, and can respond correctly in case of emergencies.

- **Software updates and patch management**: Regularly updating system software, operating systems, and other applications with the latest patches is crucial to maintain cybersecurity and system performance. It's important to validate and test software updates in a controlled environment before deploying them in the live system.

- **Compliance and audits**: Regular audits should be performed to ensure adherence to internal and external standards, regulations, and best practices. This includes safety regulations, environmental standards, and cybersecurity frameworks.

- **Documentation**: Maintaining up-to-date documentation is important for effective operation and maintenance. This includes system manuals, procedures, equipment specifications, and maintenance logs. Documentation aids in troubleshooting, training, and audits.

- **Vendor support**: Engaging with suppliers and manufacturers for their expertise and support can also be beneficial in operation and maintenance. They can provide valuable advice, technical support, spare parts, and possibly updates and upgrades.

On a daily basis, operational staff should monitor system and network activity logs to identify unusual patterns that may indicate a cybersecurity event. They should keep an eye on the status of antivirus and IDSs, ensuring they are up to date and operational. Immediate action on alerts and anomalies is crucial to prevent potential breaches from escalating.

Weekly routines might include reviewing and verifying user access logs to ensure only authorized personnel are interacting with the systems, along with checking for and applying critical security patches and updates. This is also a good cadence for testing backup processes and the integrity of stored data to prepare for potential recovery scenarios.

Monthly tasks often involve more thorough inspections and reviews. Security personnel should perform system audits to identify potential security gaps and apply necessary measures to mitigate those risks. There should also be a review of all cybersecurity policies and procedures to ensure they are current and in alignment with the latest **Threat Intelligence** (**TI**) and best practices.

Quarterly, organizations should undertake more strategic reviews and exercises. This may include penetration testing to simulate an attack on systems, training drills to assess the readiness of response teams, and a detailed audit of cybersecurity measures against compliance standards. Additionally, this could be the time to evaluate the effectiveness of training programs and update them as needed to keep pace with evolving threats.

Annually, a holistic review of the entire cybersecurity posture should be conducted. It should incorporate a comprehensive risk assessment, an audit of all cybersecurity maintenance activities, and an update of the cybersecurity IRP. It's also an opportunity to conduct an in-depth review of contracts and security measures with external vendors and partners. Annual maintenance should reexamine DRPs and BCPs to ensure they are sound and can be executed accurately in the event of a cybersecurity incident. Intensive refresher training for all relevant personnel is vital to reinforce the importance of cybersecurity in maintaining safe operations.

These frequent and structured cybersecurity maintenance activities across different timelines serve as a framework for operational vigilance. By embedding regular tasks into the routine of running ICS and SIS environments, organizations can significantly reduce the risk of cyber incidents and maintain the highest levels of safety and operational continuity.

Summary

Throughout this chapter, we reiterated the notion that ICS security isn't simply a milestone but an ongoing commitment to advancement and refinement. Adopting a comprehensive and integrated strategy is crucial for the enduring protection and reliability of any asset. We explored the importance of addressing current and relevant challenges through a robust CSMS, solid governance models, and a sustainable operational and maintenance strategy. This approach not only safeguards assets over their entire lifecycle but also ensures they keep pace with the dynamic landscape of cyber threats and technological evolution.

In *Chapter 6, Cybersecurity Risk Management of SISs*, we will explore risk management for ICS assets and, in particular, SISs.

Further reading

- *Demonstrating safety of software-dependent systems*: `https://www.dnv.com/Publications/safety-4-0-project-demonstrating-safety-of-software-dependant-systems-225267`

- *Shifting the Balance of Cybersecurity Risk: Principles and Approaches for Security-by-Design and -Default*: `https://www.cisa.gov/sites/default/files/2023-04/principles_approaches_for_security-by-design-default_508_0.pdf`

- *Secure by Design Principles*: `https://www.security.gov.uk/guidance/secure-by-design/principles/`

- *Cyber Security Procurement Language for Control Systems*: `https://www.cisa.gov/sites/default/files/2023-01/Procurement_Language_Rev4_100809_S508C.pdf`

- **Industry IoT Consortium (IIC)**: `https://www.iiconsortium.org/`

- *GDPR*: `https://gdpr-info.eu/art-25-gdpr/`

- *CISA SBOM*: `https://www.cisa.gov/sbom`

- **National Security Agency (NSA)**—*Recommendations for Software Bill of Materials (SBOM) Management*: `https://media.defense.gov/2023/Dec/14/2003359097/-1/-1/0/CSI-SCRM-SBOM-MANAGEMENT.PDF`

Part 3:
Risk Management
and Compliance

Securing **Safety Instrumented Systems (SISs)** presents unique challenges due to their critical need for high levels of uptime and reliability, combined with complex risk profiles. *Chapter 6* addresses these challenges by exploring risk and vulnerability assessment techniques tailored for SISs. It aims to help you understand and mitigate the vulnerabilities and threats associated with these mission critical systems, ensuring risks are reduced to **As Low As Reasonably Practicable (ALARP)**.

The evolving landscape of cyber threats highlights the need for robust industry legislation to protect process safety and cybersecurity systems. *Chapter 7* delves into various regulatory compliance standards, providing a comprehensive mapping of cybersecurity-related controls. This chapter draws on resources from industry bodies and centers of excellence to help organizations meet these stringent requirements.

Finally, *Chapter 8* looks ahead to the future of **Industrial Control Systems (ICSs)** and SISs, examining upcoming innovations and the challenges they will pose. This chapter discusses emerging technologies and their potential impact on the security and functionality of ICSs and SISs.

This part has the following chapters:

- *Chapter 6, Cybersecurity Risk Management of SISs*
- *Chapter 7, Security Standards and Certification*
- *Chapter 8, The Future of ICS and SIS: Innovations and Challenges*

This section equips readers with the knowledge and tools needed to understand, assess, and secure SISs in an increasingly complex cyber threat environment.

6

Cybersecurity Risk Management of SISs

This chapter provides a deep dive into the critical domain of risk assessment specifically tailored for **Safety Instrumented Systems (SISs)**. As foundational components in our **Industrial Control Systems (ICSs)**, securing SISs is paramount to the safe and uninterrupted operation of mission critical systems.

In this chapter, we will unravel the complexities of conducting a comprehensive, systematic, and methodical ICS and SIS cybersecurity risk assessment. We will highlight the significance of identifying assets, understanding potential threats and vulnerabilities, calculating impact values, and ultimately establishing the risk associated with process safety operations. This process provides vital insights that guide an organization's cybersecurity strategy and decision-making.

We also intend to illuminate the important distinction between general IT risk assessments and those specifically designed for SISs within the ICS context. This will entail dissecting unique considerations and factors that come into play with SISs, providing a clear picture of the intricacies involved in ICS cybersecurity risk assessment.

The aim of this chapter is to equip you with the knowledge and means to construct efficient and effective risk assessments for SISs, forming the backbone of your ICS cybersecurity management strategy. By the end of this chapter, you will better understand how comprehensive risk assessments contribute to enhancing cybersecurity resilience and minimizing the risk exposure of SISs.

This chapter will address the following topics:

- Importance of cyber risk assessment
- Risk assessment objectives
- SIS risk assessment principles
- Consequence-based risk assessment

- Cybersecurity risk assessment methodologies
- Conducting risk assessments in SISs
- The continuous nature of risk assessment

Importance of cyber risk assessment

A cyber-physical risk assessment is a multifaceted process that systematically evaluates and measures the potential risks linked with safety operations and workflows in our ICS landscape. These assessments provide a comprehensive, quantified understanding of vulnerabilities, potential threats, and their implications on the operating integrity and safety functions of our SIS.

This chapter underscores the imperative of risk assessments in fortifying cyber defenses, particularly in the context of SISs, which are often at the crossroads of highly sensitive industrial operations. In a cyber world fraught with evolving threats, an effective risk assessment can mean the difference between a swiftly mitigated vulnerability and a potentially debilitating intrusion into **Safety Critical Systems (SCSs)**.

Through a crisp exploration of varied risk assessment methodologies, high-level risk appraisals, and detailed risk inspections, this chapter sets the stage to foster more resilient, secure, and robust SISs. Here, in this chapter, we recalibrate our lens to view our SISs not just as functional components but as potential cyber risk-bearing entities – a perspective that could drive powerful preventive and protective action against cyber threats.

This foundation should provide a comprehensive journey through the multidimensional topic of risk assessment for SIS cybersecurity, equipping readers with the knowledge they need to conduct effective, meaningful assessments in their own facilities.

As automation and digitalization pervade the industrial sector, ICSs and SISs have become the lifeline of critical infrastructures that underpin key aspects of modern life, spanning industries such as energy production, water treatment, manufacturing, and transportation. Protecting these systems from cyber threats has metamorphosed from an IT concern into a crucial, nationally imperative mission.

At the heart of this protective endeavor lies a process that remains pivotal to cyber-defense architectures – risk assessment. Shaped by a systematic methodology, cybersecurity risk assessment is an effective tool for identifying, analyzing, and evaluating potential risks that could threaten the cybersecurity architecture of ICSs and impede the functional integrity of SISs.

For SISs, which are specifically designed to marshal industrial processes to a safe state during anomalous scenarios, cybersecurity risk assessments are particularly vital. On the one hand, SISs provide an extra layer of protection against hazardous events. They enforce safety functions when process conditions exceed predefined parameters, ensuring that any dangerous deviation is mitigated before it escalates to a full-blown safety incident.

On the other hand, their critical position within the **Process Control Network (PCN)** and increasingly entwined communication with **Basic Process Control Systems (BPCSs)** make an SIS a potential target for cyber threats. A successful breach into the SIS could offer a malicious threat actor the ability to manipulate the system, potentially inducing false trips or, worse, suppressing the fulfillment of a safety function during a genuine demand scenario. It's a cyber risk that could have catastrophic physical consequences, turning an industrial facility into a hazard zone.

Here's where cybersecurity risk assessments step in. They serve as a compass, illuminating the path forward by identifying potential vulnerabilities inherent in the system, estimating the likelihood of a cyber threat exploiting these vulnerabilities, and evaluating the impact such an incident could have on safety, environment, and system operability. By performing a risk assessment, organizations can focus on the security landscape's *big picture*, comprehensively visualizing all possible entry points a potential cyber threat could leverage and the specific severity associated with each vulnerability.

Robust cybersecurity risk assessments provide organizations with valuable insights necessary to make informed decisions about security investments and resource allocation. They offer the rationale for implementing protective measures, designing emergency response protocols, establishing **Security Operations Centers (SOCs)**, and maintaining a solid cybersecurity posture.

The following diagram summarizes some business and technical key areas of cyber risk assessment in the context of ICSs:

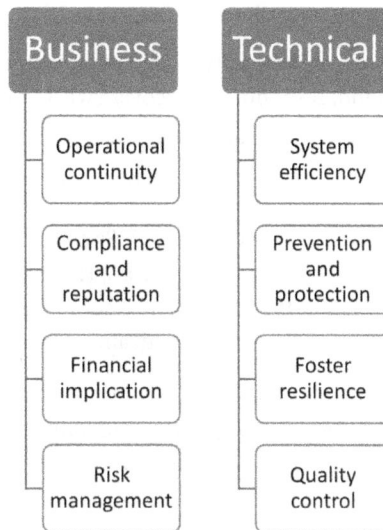

Figure 6.1 – Cyber risk assessment key areas

Here are some key business areas and their implications:

- **Operational continuity**: A cyber attack on an SIS could potentially create hazards and disrupt business operations, leading to financial loss or ceasing the license to operate. A risk assessment allows businesses to protect themselves from such eventualities.

- **Compliance and reputation**: Regulations require that companies take adequate measures to mitigate cyber risks. Non-compliance may result not only in legal penalties but also in loss of reputation, which can impact a firm's market position.

- **Financial implication**: Understanding and mitigating cyber risks can prevent significant financial losses from incidents such as data breaches or system downtime. The cost of a risk assessment is small in comparison to potential losses.

- **Risk management**: From a business standpoint, understanding cyber risks allows decision-makers to prioritize resources effectively and manage risks optimally.

A breakdown of the technical perspective is listed as follows:

- **System efficiency**: An SIS risk assessment can help identify system vulnerabilities and inefficiencies, enabling their correction and thus improving overall system performance

- **Prevention and protection**: Risk assessments allow businesses to identify, address, and prevent potential threats before they materialize, thus ensuring the functionality and lifespan of the system

- **Foster resilience**: From a technical perspective, risk assessments also allow engineers to design resilient SISs that can withstand, respond, and recover swiftly from cyber attacks

- **Quality control**: Regular risk assessments ensure that safety and security standards are maintained and updated, contributing to overall quality control in system design and operation

Moreover, cybersecurity risk assessments in the context of SISs and ICSs offer businesses a myriad of benefits. They provide a structured approach to managing complex security risks, comply with various regulatory and standards requirements, allow for effective and efficient prioritization of security resources, and ultimately ensure the security and safety of the organization's people, environment, and assets.

Notably, risk assessments are not a one-off exercise but a continuous process, mirroring the evolving landscape of cyber threats. As technology evolves, new vulnerabilities may emerge, and previously non-significant threats could grow in relevance. Regular cybersecurity risk assessments allow organizations to stay agile, continually upgrading their defenses and minimizing cyber risk exposure.

Cybersecurity risk assessments are integral to formulating a robust security strategy for protecting essential processes controlled by the ICS and, particularly, the SIS. As we increasingly rely on these systems to regulate critical industrial operations safely, cybersecurity intrinsic to SIS resilience is more than a mere operational necessity. It's an essential duty to our employees, communities, environment, and the national infrastructure, lending credibility and confidence in the operational continuity of our industrial setups amid burgeoning cyber risks.

After emphasis on cybersecurity risk assessment, let's examine various elements of risk assessment in the context of SISs.

Risk assessment objectives

The paramount objective of conducting a cybersecurity risk assessment in an ICS environment, especially related to safety critical components such as SISs, is to gain a comprehensive understanding of potential security risks lurking within the system. Achieving this objective is reliant on several key outcomes.

The following table breaks down the main objectives of an ICS cyber risk assessment:

Objective	Description
Identify vulnerabilities	This involves reviewing secure network design and architecture, operation, and maintenance of safety systems to identify potential weaknesses. This involves identifying any elements of the SIS that are susceptible to cyber threats.
Secure network architecture	
Effectively manage risks	The assessment should provide insights into how to manage cyber risks effectively. This involves developing and implementing strategies and procedures to mitigate identified risks.
Augment safety measures	The cyber risk assessment should help strengthen existing safety measures. This is accomplished by providing recommendations on how to enhance security protocols and safety barriers.
Compliance with regulations	The assessment ensures the SIS is compliant with relevant industry standards and regulations pertaining to cybersecurity.
Continual improvement	The assessment should provide a benchmark for future assessments, helping with the measurement of improvement over time.
Raise awareness	Another objective of the assessment is to bring attention to potential cyber threats. It aims to educate relevant personnel about cybersecurity and the ways in which SISs could be potentially breached.
Incident management (IM)	The assessment can help in planning for a cyber incident, including identifying potential threats, risks, the likely impacts of different types of incidents, and appropriate responses.

Table 6.1 – ICS cyber risk assessment main objectives

To start off, there is a crucial need to identify and determine specific business requirements and drivers that govern the ICS environment. This understanding paves the way for efficient and target-focused risk assessment. Next, it's important to detail specific ICS components that could potentially be subject to cyber threats and risks. This intense scrutiny helps sharpen our focus on areas critical to maintaining the integrity of the entire system.

To further strengthen the risk assessment, an identification of cybersecurity weaknesses and vulnerabilities within the system is necessary. This step uncovers potential openings that could give threat actors access to the system. Going a step further, we also aim to identify specific targets or vulnerabilities that hold the potential of being exploited by such cyber threats.

Critical to this comprehensive approach is the identification of threat actors and dissecting their motivation, capability, and intent with regard to our system. This exercise offers an insight into the *who* and *why* behind potential attacks, allowing for more nuanced risk mitigation planning.

There's also a strong emphasis on determining the most *credible consequences*, based on inherent risk evaluation. These credible consequences essentially define the business impact that could occur if a cybersecurity threat becomes a reality. Understanding this helps us gauge the severity of the potential risk and strategize accordingly.

Finally, we aim to establish risk ratings that serve as the backbone for determining **Safety Integrity Levels** (**SILs**) for functional groups of systems. This classification helps prioritize areas needing urgent attention and fosters an all-encompassing approach to ensure security in the ICS environment. Thus, the widespread aim of an ICS risk assessment is to build a fortified shield of cybersecurity measures that are tailored to fit the unique requirements and potential risks of a given ICS environment.

In the context of an ICS environment, the cyber risk assessment scope encompasses the examination of every facet of the ICS infrastructure for potential cyber vulnerabilities and threats. This ranges from operator workstations and portable devices to **Programmable Logic Controllers** (**PLCs**), **Remote Terminal Units** (**RTUs**), servers, networks, and even the human factors involved, such as training and awareness.

The assessment focuses on all relevant (known) types of threats, including both internal and external, deliberate or accidental, ranging from malware to insider threats. It also covers both physical and digital risks, including hardware vulnerabilities, data transmission and storage security, software flaws, and network vulnerabilities. The consequent impact of potential attacks on these vulnerabilities, such as disruption of business processes, loss or alteration of critical data, damage to physical equipment, or even catastrophic failures leading to injury, loss of life, or environmental damage, is also part of the evaluation process. The final objective is not only to identify these vulnerabilities and risks but also to determine feasible mitigation strategies and response plans to enhance the overall cyber resilience of the ICS environment.

To ensure dependable and consistent security for SISs, the process for evaluating cybersecurity risks should be implemented through the interplay of the following:

- The SIS in question
- The methodology employed for the cybersecurity risk assessment
- The regularity with which the SIS risk assessment is performed
- The expertise of the team performing the assessment
- Tools and resources (for example, checklist of actions, inventory catalog, network documentation)
- The quality and time span of the SIS risk assessment
- Documentation and verification of the results
- Management of identified vulnerabilities

The SIS risk assessment can also be conducted following methods different from this one. However, the principles and guidelines of this method and the previously mentioned standards should be considered.

We will deep dive later into examples of scope based on industry standards and best practices. Prior to this, let's first examine the concepts of risk assessment, especially for SISs.

SIS risk assessment principles

Understanding cybersecurity risk within ICSs and SISs involves acknowledging the interplay between threats, vulnerabilities, impacts, and consequences. A threat represents a potential source of harmful actions that could affect system operations, arising from various sources such as natural disasters, malicious attacks, or human errors. Vulnerabilities are weaknesses in a system that can be exploited by threats, potentially disrupting normal operations. Impacts represent the consequences of these threats exploiting vulnerabilities, which can range from minor system disruptions to significant safety incidences.

Risk assessment plays a pivotal role in a **Cybersecurity Management System (CSMS)**. It provides a basis for understanding cyber risks associated with an organization's ICS, including the SIS. Through identification, analysis, and evaluation, risk assessments help to devise strategies that efficiently mitigate these risks, protecting the integrity of the ICS and the critical infrastructures they support.

However, it's crucial to differentiate between general IT risk assessments and ICS-/SIS-specific risk assessments. While both aim to identify threats, vulnerabilities, and impacts, the priorities and context of operations differ greatly. IT systems prioritize the confidentiality and integrity of data, whereas SISs are built to ensure industrial process safety, targeting primarily the availability of systems. The inherent differences in their usage, design, and operational methodology require separate approaches to risk assessment. An SIS-specific risk assessment considers its unique vulnerabilities, threats, impacts, and the resulting costs of functional failure, making it a critical tool in maintaining the security and safety performance of these systems.

In addition, if we explore this further, then it becomes clear that SIS risk assessments, from both a functional safety and cybersecurity perspective, often involve the identification of hazards/threats and the assessment of their impact on the overall system. However, the focus and methodologies adopted can vary.

From a functional safety perspective, the SIS risk assessment mainly focuses on the reliability and safety of the SIS in preventing or mitigating hazardous events. It involves evaluating the system's performance levels, safety features, and fail-safe designs. It considers physical threats, such as component failures, process upsets, environmental conditions, and human errors. A functional safety risk assessment uses standards such as *IEC 61511* and *IEC 61508*, which provide guidelines for ensuring the safety integrity of safety systems.

On the other hand, an SIS risk assessment from a cybersecurity perspective is concerned with the protection of the system against cyber threats. It focuses on the confidentiality, integrity, and availability of SIS data and services. It considers threats such as software bugs, network vulnerabilities, and potential cyber attacks that could compromise the system. A cybersecurity risk assessment uses standards such as *IEC 62443*, which offers guidelines for the security of **Industrial Automation and Control Systems (IACSs)**.

In terms of differences, functional safety does not usually consider deliberate attacks and is more focused on systematic failures or random faults. In contrast, cybersecurity considers both accidental and deliberate threats, including malicious attacks.

Common areas between functional safety and cybersecurity include the necessity to maintain system reliability and protect against potential threats. Both require an understanding of the system, processes, and potential risks. Additionally, both perspectives require the development of risk-reduction measures and response strategies.

Finally, while both types of assessments have their unique foci, it's crucial that they are conducted in a coordinated manner. The increasing interconnectivity and digitalization of systems have blurred the boundaries between functional safety and cybersecurity, making it vital for organizations to integrate these perspectives to maintain the safety and reliability of their SIS efficiently.

Consequence-based risk assessment

The consequence-based risk assessment approach for ICSs and SISs is centered on understanding and managing potential outcomes or impacts that could result from a security compromise.

Unlike traditional IT risk assessments, which often focus on the likelihood of a threat, the consequence-based approach starts by identifying worst-case scenarios or major impacts, such as environmental harm, equipment damage, financial loss, or even a threat to human safety. Based on the identified consequences, security measures are then tailored to prevent or mitigate these specific impacts.

In essence, this approach is grounded in the principle that the higher the potential consequences, the more rigorous the security controls should be. This methodology is particularly effective in the context of ICSs and SISs where system availability, functionality, and integrity often take precedence over other traditionally IT-focused considerations, such as confidentiality.

The following table highlights the principal differences between the consequence-based risk assessment that is typically performed in the context of ICSs and SISs and the traditional IT risk assessment utilized for data security:

Aspect	Consequence-based Approach	Traditional IT Risk Assessment
Primary goal	Ensuring system functionality and physical safety	Safeguarding data confidentiality and integrity
Risk prioritization	Prioritize risks based on potential operational, safety, and environmental consequences of system disruptions	Prioritize risks based on the likelihood of occurrence
Security focus	Focuses on the system's availability, ensuring its ability to maintain continuous operations	Focuses on data protection from data breaches
Acceptable downtime	Zero tolerance for outages to prevent harmful consequences	Some tolerable brief outages
Threat perception	Emphasizes more the consequences of physical damage or harm, productivity loss, and potential safety threats	Concentrates primarily on cyber threats and data loss
Risk mitigation strategy	Design safety measures to prevent system disruptions that can lead to severe impacts	Implement strategies mainly to counteract data breaches or hacking

Table 6.2 – Consequence-based risk assessment aspects

Risk criteria in the context of ICS cybersecurity risk assessment are standards or guidelines that an organization establishes to measure and tackle risks facing their ICS infrastructure. These criteria reflect the organization's risk appetite or the level of risk it is prepared to accept before action is deemed necessary. Corporate risk criteria in ICS cybersecurity consider factors such as potential impacts on operational performance, safety of personnel, regulatory compliance, and unintentional system downtime. They also look at potential financial implications, such as costs of mitigating risks, potential fines for non-compliance, and liability or loss potential associated with cyber attacks. Companies often establish thresholds for each of these elements, defining acceptable levels and when triggers for action should be set.

A proper understanding and application of these risk criteria is critical as it guides the risk assessment process, helping to prioritize risks and formulate effective cybersecurity strategies to ensure the availability, reliability, integrity, and safety of ICSs.

An example of typical risk elements used within an organization in the process industry is listed as follows:

Severity	Consequences			
	People	Assets	Environment	Reputation
0	No injury or health effect	No damage	No effect	No impact
1	Slight injury or health effect	Slight damage <US $10,000	Slight effect	Slight impact
2	Minor injury or health effect	Minor damage between US $10,000 and US $100,000	Minor effect	Minor impact
3	Major injury or health effect	Moderate damage between US $100,000 and US $1 million	Moderate effect	Moderate impact
4	Permanent disability or up to three fatalities	Major damage between US $1 million and US $10 million	Major effect	Major impact
5	More than three fatalities	Massive damage > $10 million	Massive effect	Massive impact

Table 6.3 – Example of severity levels

In addition to the severity level, the majority of process industries assess their risk profile based on the likelihood and consequences of potential events, as illustrated in the following table:

Consequences	Likelihood				
	A	B	C	D	E
0	Light Blue – Low	Light Blue – Low	Light Blue – Low	Light Blue – Low	Light Blue – Low
1	Light Blue – Low	Light Blue – Low	Dark Blue – Medium	Dark Blue – Medium	Dark Blue – Medium
2	Light Blue – Low	Dark Blue – Medium	Dark Blue – Medium	Yellow – High	Yellow – High
3	Dark Blue – Medium	Dark Blue – Medium	Yellow – High	Yellow – High	Red – Serious
4	Dark Blue – Medium	Yellow – High	Yellow – High	Red – Serious	Red – Serious
5	Yellow – High	Yellow – High	Red – Serious	Red – Serious	Red – Serious

Table 6.4 – Example of risk ranking matrix

These levels can be adapted according to the specific needs and risk tolerance of your organization. The likelihood levels should be based on a combination of historical data, predictive analytics, expert opinions, and the nature and complexity of your SIS.

Besides risk criteria, risk ranking plays a vital role in ICS cybersecurity risk assessment as it provides a methodical way to prioritize cybersecurity risks based on their potential impact and probability of occurrence. In the complex environment of ICSs, threats can come from multiple sources, including hardware malfunctions, software bugs, human errors, or malicious cyber attacks. A properly conducted risk ranking helps organizations identify which risks pose the most significant threat to the operation and safety of their control systems. Factors taken into account can include the severity of potential consequences, the vulnerability of the system to a particular threat, and the likelihood of that threat materializing. The outcomes of this ranking process drive the allocation of resources, guiding the development of targeted mitigation strategies and aiding in making informed decisions on where to focus preventive and protective efforts.

To help prioritize and justify risk decisions, **As Low As Reasonably Practicable (ALARP)** is a safety concept that seeks to reduce risks to a level that is acceptably low, considering the time, effort, and resources necessary to manage the risk further. According to the ALARP principle, if a risk is identified within the SIS, efforts should be made to mitigate it to a level that is as low as is reasonably practicable. In the context of SIS cybersecurity risk assessment, this can involve implementing safeguard measures, monitoring systems, **Incident Response (IR)** planning, staff training, and other mitigative actions.

Similarly, the **As Low As Reasonably Achievable (ALARA)** principle emphasizes reducing risks to the lowest level possible, taking into account economic and societal factors. ALARA is particularly relevant in contexts where exposure to hazards, such as radiation, must be minimized. As with ALARP, ALARA involves continuous monitoring, implementing protective measures, and ensuring that any risk-reduction actions are proportional to the benefits achieved.

The key determinant in deciding what is reasonably practicable or achievable is proportionality; the time, effort, and resources spent on mitigating risks must be proportional to the reduction in risk achieved. Therefore, if a risk is minor and the measures to reduce it are disproportionate in terms of cost and effort, then it could be deemed ALARP or ALARA without further mitigation. Conversely, if a risk could have significant implications but can be reduced by proportionate measures, then those measures should be taken. However, all risks, regardless of their ALARP or ALARA status, should be regularly reviewed and reassessed for changes in circumstances or new information that might affect their status.

As depicted in *Figure 6.2*, ALARP and ALARA principles are widely applied to delineate acceptable risk levels in various processes, categorizing risks into intolerable, tolerable, and broadly accepted classifications:

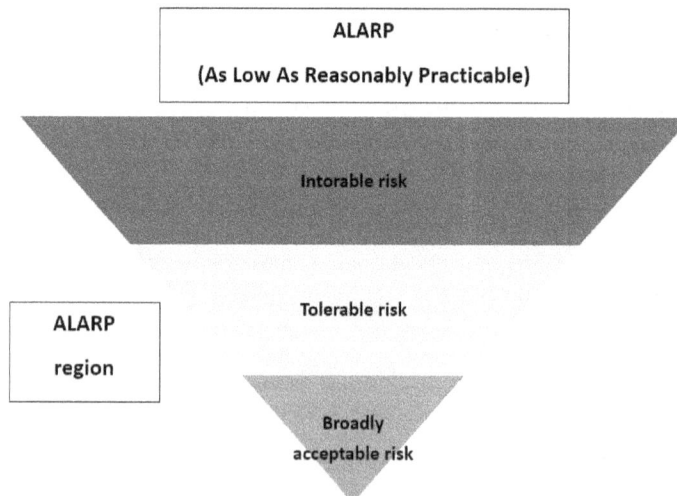

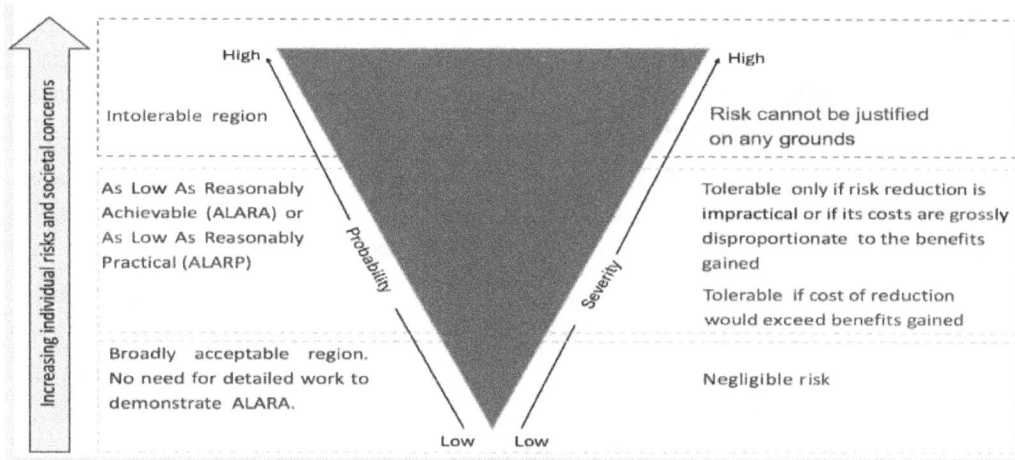

Figure 6.2 – ALARP and ALARA

We defined the foundation of risk assessment with a focus on consequences and SIS specification. Now, let's see this process in action based on industry best practices.

Cybersecurity risk assessment methodologies

There are several methodologies available for carrying out risk assessments for ICSs, each with its own strengths and weaknesses. In this section, we will focus only on methodologies and standards that are SIS-related.

Starting with *IEC 62443*, this stands as an international standard for IACS security. It provides a flexible framework to address and mitigate current and future security vulnerabilities in ICSs. Its strength lies in its comprehensive approach to ICS security, covering aspects of policy, system design, and procedural controls. However, a detailed risk assessment in line with *IEC 62443* can be complex and resource-intensive due to its broad scope and depth.

The **National Institute of Standards and Technology (NIST) Special Publication (SP) 800-82** standard provides guidance specifically tailored to ICSs. It is comprehensive and provides strong procedures for addressing the ICS lifecycle, but implementation can often be complex and time-intensive due to the granular level of detail in the recommendations.

BowTie is a risk evaluation method that offers a visual depiction of causal relationships in high-risk scenarios. Its strength lies in its simplicity and ease in identifying preventive and mitigative controls. However, it might not delve into the depths of detailed analysis as it provides a broader, more visual-oriented overview of risks.

NAMUR is a user-friendly approach, which makes it accessible to a range of users in the ICS landscape. It focuses on functional safety and presents a clear guide to the main phases of life cycle management. The downside to NAMUR is its lack of direct reference to cybersecurity, as it was primarily designed with process safety management in mind.

All in all, the appropriate methodology largely depends on specific requirements, complexity, and existing risks within an organization's ICS landscape. It's strongly advised to integrate a combination of methodologies to achieve a robust and comprehensive **Risk Management Plan** (**RMP**).

Before we dig into a specification of some of these methodologies and standards, we would like to provide a high-level mapping and analysis of each of these in the following table:

Methodology	Benefits	Limitations
NAMUR	• Provides a well-defined quantitative structure for risk assessment • Offers detailed analysis on uncertainty and mitigation randomness • Clear coverage of safety and availability	• Quite a new approach; limited real-world implementation examples • May require significant effort and time for multiple assessments
IEC 62443-3-2	• Comprehensive and well-established for IACSs • Provides a systematic and practical approach to identifying and addressing security vulnerabilities|	• Implementation can be complex and time-consuming • Some processes may not be suitable for all industries
BowTie	• Visual representation aids comprehension • Can accommodate complex risk scenarios • Effective in communicating risk situations to various stakeholders	• Relatively simplistic; might not capture all details • Requires extensive knowledge and data for the setup
NIST 800-82	• Covers an extensive list of security controls • Provides detailed guidance on implementation	• Primarily designed for the US market • May require a significant amount of time to fully implement

Table 6.5 – Methodologies analysis

The selection of a particular methodology should take into account the specific needs, resources, regulatory requirements, and risk tolerance of your organization. It's also worth considering the possibility of applying a hybrid approach that combines favorable aspects of different methodologies to best suit the organization's objectives.

Conducting risk assessments in SISs

For the context of SISs, we will highlight how a risk assessment is conducted with a focus on *IEC 62443*, NAMUR, and BowTie as widely adopted standards in process industries.

IEC 62443-3-2

The *IEC 62443-3-2* standard utilizes a qualitative approach to cyber risk assessment aimed at identifying the potential for cyber attacks that could lead to incidents harming human life, the environment, property, or operational capabilities. A detailed summary of this methodology is provided in the following diagram:

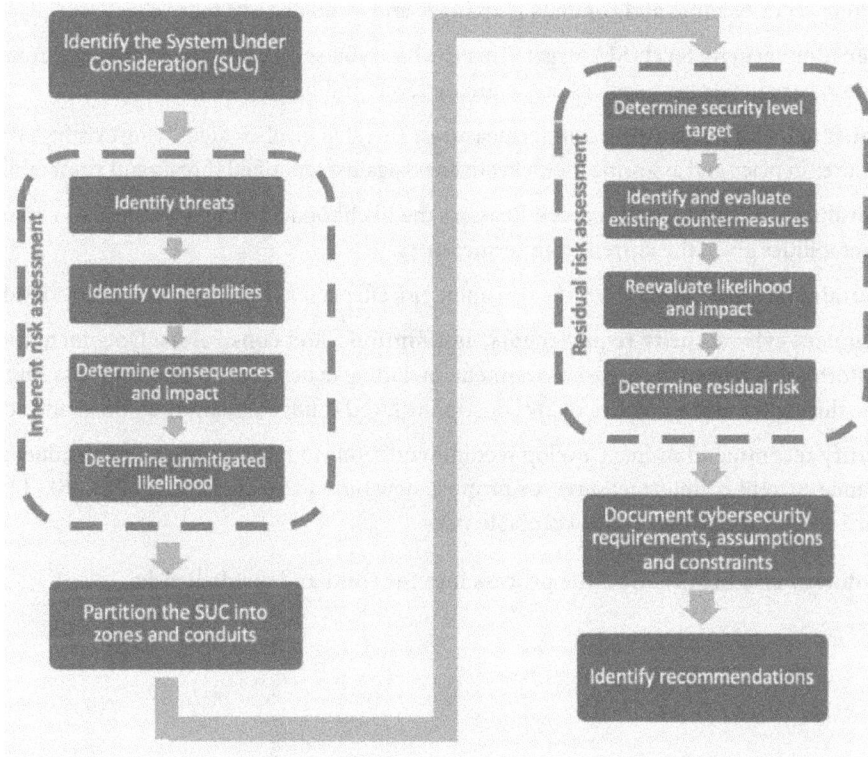

Figure 6.3 – Diagram depicting the HAZOP method for risk
assessment based on the IEC 62443-3-2 standard

As per *Figure 6.3*, the steps taken for the high-level risk assessment were the following:

1. **Identify/select systems under consideration (SuCs)**: This step is about defining the scope of the risk assessment. It involves identifying SISs and other ICSs that will be assessed.

2. **Identify threats and vulnerabilities**: This involves identifying possible cybersecurity threats and detecting vulnerabilities within the specified systems. Threats could be external, such as malicious attacks, or internal, such as system bugs or human errors.

3. **Determine consequences and impact**: This step requires understanding the potential repercussions if threats were to exploit identified vulnerabilities. Consequences may include system downtime, data breaches, or safety incidents.

4. **Determine unmitigated likelihood**: This involves assessing the probability of identified threats considering existing vulnerabilities, but without considering any countermeasures currently in place.

5. **Determine inherent risk**: Inherent risk is the level of risk present in the absence of countermeasures.

6. **Partition the SuC into zones and conduits**: This step involves segmenting the system into different security zones and conduits to manage and evaluate risks more effectively.

7. **Determine security level (SL) target**: This is where you set the desired SL for each zone and conduit, based on factors such as consequence levels and business requirements.

8. **Identify and evaluate existing countermeasures**: This step involves identifying existing security measures in place and assessing their effectiveness against identified threats and vulnerabilities.

9. **Re-evaluate likelihood and impact**: Reassess the likelihood and impact of threats exploiting vulnerabilities given the current countermeasures.

10. **Determine residual risk**: Evaluate the remaining risk after countermeasures have been considered.

11. **Document cybersecurity requirements, assumptions, and constraints**: Documentation of all information related to the risk assessment, including cybersecurity needs, any assumptions made during the assessment, and any constraints faced (budget, manpower, time, and so on).

12. **Identify recommendations**: Develop recommendations to mitigate identified residual risks, enhance current countermeasures, or propose new ones to reach the **target SL** (**SL-T**) with regard to identified threats and acceptable risks.

The methodology also breaks down the process into inherent and residual risks.

Inherent risk assessment

Inherent risk assessment focuses primarily on identifying threats and vulnerabilities:

- **Threat identification**: Within the **operational technology** (**OT**) domain, potential threats manifest in various forms, such as unauthorized reprogramming of control devices, manipulation of control logic, **denial-of-control** (**DoC**) actions, falsification of system status information, and possible modification of SISs. This section outlines pertinent threat sources and delineates cybersecurity threats specifically applicable to the OT environment. It is advisable to incorporate **threat intelligence** (**TI**) from governmental sources and sector-specific information-sharing and analysis centers, if available, for a comprehensive threat assessment.

- **Threat agents**: The probability of a cybersecurity event occurring hinges upon the amalgamation of threat actors' capabilities, motivations, and opportunities. Opportunity, in turn, is contingent upon the presence of vulnerabilities. A comprehensive list of potential threat actors targeting IACSs is available in NIST *SP 800-82, Revision 2, Guide to Industrial Control Systems (ICS) Security*, specifically detailed in *Reference/A9/Appendix C*.

- **Threat scenarios**: The landscape of threats is dynamic, necessitating continual assessment to accurately depict relevant threat scenarios. A robust threat description encompasses the following:

  - Identifying the threat source

  - Detailing its capabilities, motivations, and opportunities

  - Outlining potential threat vectors

  - Identifying assets susceptible to these threats. *Reference/A9/Appendix C, Table C8* in *NIST SP 800-82, Revision 2, Guide to Industrial Control Systems (ICS) Security* provides an inclusive list of potential threat scenarios for further analysis.

- **Vulnerabilities identification**: Vulnerabilities represent weaknesses within IACSs and associated protocols that could potentially be exploited by threat actors. The outcome of this phase should culminate in a comprehensive inventory of vulnerabilities, each aligned with corresponding threats. Prior to concluding this process, it is advisable to cross-reference the list delineated in *NIST SP 800-82, Revision 2, Guide to Industrial Control Systems (ICS) Security* (*Reference/A9/Appendix C, Tables C2 to C7*) to ensure no pertinent vulnerabilities have been overlooked.

- **SIS threats and vulnerabilities identification**: The *IEC 61511* standard (*Reference/A8*) directs attention to *ISA TR 84.00.09, Security Countermeasures Related to Safety Instrumented Systems (SIS)* (*Reference/A10*). To adhere to the risk assessment stipulations outlined in this standard, it is imperative to incorporate hazards and threat agents pertinent to SIS, as elucidated in *Chapter 5*, into the risk assessment process. The company mandates completion of the NAMUR *NA 163, Security Risk Assessment of SIS* (*Reference/A11*) checklist for their SISs to support comprehensive risk assessment efforts. This entails determining the consequences, impact, and unmitigated likelihood of potential risks.

- **Consequences and impact determination**: An assessment will be conducted for each identified threat and vulnerability to ascertain potential consequences and impact in the event of their exploitation. It is advisable to articulate the worst-case-scenario outcome of the outlined threat scenarios across key risk domains, encompassing personnel safety, business repercussions (including production loss and damage), reputation tarnishing, and environmental impact. The outcome of this phase will yield a comprehensive inventory detailing the consequences and impact of each threat and vulnerability should countermeasures not be implemented.

- **Initial likelihood determination**: Evaluate the initial likelihood of each scenario, excluding consideration of existing cybersecurity countermeasures. During this assessment, take into account non-cyber **Independent Protection Layers (IPLs)**, such as physical security measures or mechanical barriers (for example, pressure safety valves), that are implemented to mitigate threats. The determination of likelihood should be qualitative, taking into consideration factors such as the capability, motivation, and opportunity of threat agents. It's important to note that opportunity is contingent upon vulnerabilities.

Determine initial cybersecurity risk based on the following elements:

- **Initial cybersecurity risk determination**: Estimate the initial cybersecurity risk associated with each identified threat and vulnerability combination by integrating measures of consequence and impact with the initial likelihood assessment.

- **Prioritize risks**: Risks are prioritized based on the levels determined, the risk score ranking, and the required risk mitigation requirements.

- **Zones and conduits**: Before conducting prioritization of a **Process Control System (PCS)** or a risk assessment, it is important that there is a clear understanding of the scope/boundaries of the systems to be assessed. A diagram depicting zones and conduits is a tool to help visualize the relation between SuCs and aid in performing the risk assessment. Mapping zones and conduits helps in generating an overview of the environment of SuCs and their relations. Having this overview allows for targeted initiatives in assigning, assessing, and redesigning security controls that are grouped into **Foundational Requirements (FRs)** (as per *IEC 62443-3-3*).

Zone and conduits have been covered in *Chapter 3* in depth.

Residual risk assessment

Residual risk assessment involves evaluating and mitigating any remaining cybersecurity risks after implementing protective measures. The process includes the steps detailed next.

Determining the SL (target and achieved)

Each security sub-zone is assigned an SL-T based on the criticality (inherent risk ranking) of assets and is defined using a risk-level criticality rating that has been aligned to the contextual business risk model ranging from *SL 1* (Low) to *SL 4* (Very High), in respect to the asset's criticality in accordance with the risk assessment results.

SL in the context of the *IEC 62443* standard refers to a discrete implementation of cybersecurity countermeasures that corresponds to a set of requirements. The aim of these structures is to protect IACSs from a specific set of threats.

The *IEC 62443* standard defines four SLs, each with a corresponding set of requirements for the system.

The associated four SLs are defined as follows:

- SL 1 aims to thwart unauthorized information disclosure through eavesdropping or inadvertent exposure

- SL 2 endeavors to prevent unauthorized information disclosure to entities employing rudimentary methods with limited resources, basic skills, and minimal motivation

- SL 3 seeks to forestall unauthorized information disclosure to entities employing advanced techniques with moderate resources, specialized IACS skills, and moderate motivation

- SL 4 strives to prevent unauthorized information disclosure to entities employing sophisticated techniques with ample resources, specialized IACS skills, and high motivation

Source: *IEC 62443-3-3* (`https://webstore.iec.ch/preview/info_iec62443-3-3%7Bed1.0%7Db.pdf`)

Each SL addresses a certain level of threat, increasing in sophistication and impact, with corresponding countermeasures to help mitigate these threats. The appropriate SL for a specific system typically depends on factors such as the potential impact of an incident, specific threats that need to be guarded against, and the risk tolerance of the organization.

Baseline control requirements are referred to as **System Requirements (SRs)**, which are applicable to all SuCs that are located in that particular security zone. SRs are grouped by seven FRs, derived from *IEC 62443 Part 3-3: System Security*.

Regarding the requirements and SLs, they encompass the following components:

- **Identification and Authentication Control (IAC)**
- **Use Control (UC)**
- **System Integrity (SI)**
- **Data Confidentiality (DC)**
- **Restricted Data Flow (RDF)**
- **Timely Response to Events (TRE)**
- **Resource Availability (RA)**

The FRs are linked to achieving a specified SL through the integration of people, processes, and technology. Each FR is aligned with a security control objective, serving as a method to conceptualize each risk mitigation strategy. These are further detailed in the following table:

ID	FR	Control Objective
FR1	IAC	The SUC shall provide the necessary capabilities to reliably identify and authenticate all users (humans, software processes and devices) attempting to access the SUC.
FR2	UC	The SUC shall provide the necessary capabilities to enforce the assigned privileges of an authenticated user (humans, software processes and devices) to perform the requested action on the system or assets and monitor the use of privileges.
FR3	SI	The SUC shall provide the necessary capabilities to ensure the integrity of the SUC to prevent unauthorized manipulation.
FR4	DC	The SUC shall provide the necessary capabilities to ensure the confidentiality of information on communication channels and in data repositories to prevent unauthorized disclosure.
FR5	RDF	The SUC shall provide the necessary capabilities to segment the control system via zones and conduits to limit the unnecessary flow of data.
FR6	TRE	The SUC shall provide the necessary capabilities to respond to security violations by notifying the proposer authority, reporting needed evidence of the violation and taking timely corrective action when incidents are discovered.
FR7	RA	The SUC shall provide the necessary capabilities to ensure the availability of the control systems against the degradation or denial of essential services.

Figure 6.4 – FRs

Source: *IEC 62443-3-3* (`https://webstore.iec.ch/preview/info_iec62443-3-3%7Bed1.0%7Db.pdf`)

The aforementioned FRs will provide guidance for the required SL-T to be achieved via the application of SRs (security services). Depending on the SL of a system, each of the FRs needs to be met with different measures. Measures for an SL-4 system will ultimately be stricter than measures for an SL-2 system. Further information can be found in *IEC 62443-3-3, Annex B*.

The currently **achieved SLs (SL-As)** per SuC will need to be determined as part of the follow-up detailed risk assessments. The difference between the required SL(SL-T) and the SL-A is considered to be the residual risk:

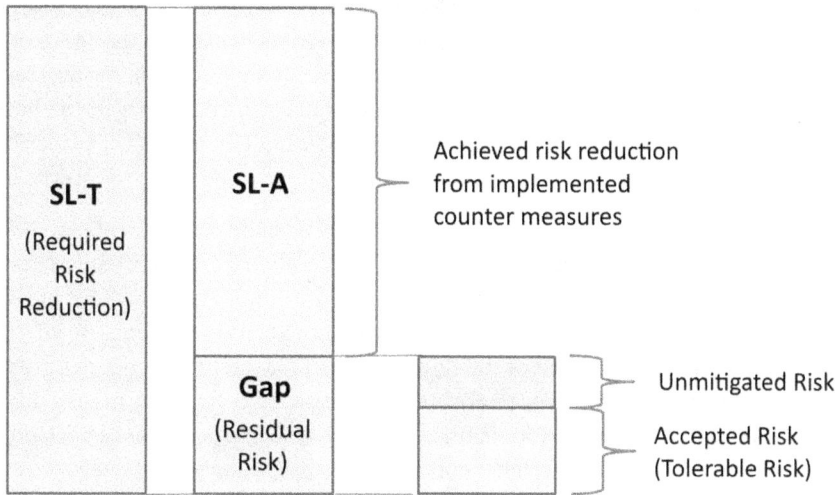

Figure 6.5 – SL-T and SL-A

Existing countermeasures identification (SL-A) and evaluation

It is essential to identify and assess existing countermeasures to gauge their efficacy in mitigating risks and reducing the likelihood or impact of potential threats, ultimately determining residual risk.

Based on the residual risk, a set of recommendations and countermeasures will be generated to effectively mitigate identified cybersecurity risks to a level that is deemed acceptable.

BowTie

The engineering field has standard procedures for evaluating process safety risks; however, applying this approach to cyber hazards in ICSs does not provide an accurate picture since cyber dangers are not foreseeable given the constantly shifting nature of threats. For assessing cyber risks, both IT and engineering expertise are necessary, and furthermore, these groups must be able to communicate effectively in order to effectively manage dangers and create deterrence. To this end, the BowTie model is widely adopted within process industries and usually used as a process safety risk assessment tool, provides a way forward; utilizing threat scenarios, it determines the effectiveness of barriers in curtailing the occurrence of a top-level event and the magnitude of damage if it does occur.

The BowTie model can be used to evaluate potential hazards resulting in various top events. A top event is acknowledged as the *first cause* of all possible consequences; however, this does not always lead to an incident. To reduce the chance of the top event occurring in the first place, as well as the impact of it, barriers must be established. A prime example of this is driving: if a vehicle loses control on icy roads, the severity of the consequence is reliant on the barriers in place. In this instance, winter tires reduce the chances of losing control, while winter driver training is a useful strategy to correct the vehicle and avoid an incident. When applying this concept to cyber risk management, businesses must consider any potential top events based on the threat landscape. Out of this finite number, a risk model can be developed that engages the security controls necessary to bring risks to an acceptable level.

The analogy between BowTie model safety and cybersecurity is readily apparent. In its most basic form, the risk management cycle in cybersecurity consists of identifying an attack vector (threat) and potential damages associated with it (consequences). As such, the BowTie model can help to visualize and understand potential attack vectors, dangers, and how individual components are affecting the security of the system. The main benefit of the BowTie model for safety and cybersecurity is that it allows teams to quickly and effectively analyze and understand complex systems. This model encourages analyses of various safety and security issues at the same time, and this can help to inform a comprehensive and effective risk management process.

The BowTie methodology offers a structured approach to risk assessment, involving the identification of pertinent threats, corresponding defenses, potential impacts, and strategies for minimizing those impacts in response to hazardous events. Within the realm of IACS security, this approach can be likened to a strategy of *preventing unauthorized access* with measures on the left side of the diagram and *mitigating the consequences in case of breach* with measures on the right side:

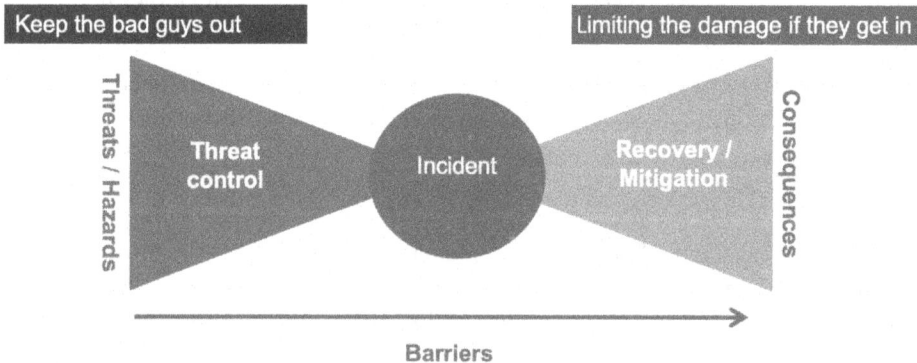

Figure 6.6 – Example of BowTie's barriers

The BowTie methodology serves as a risk assessment technique offering a straightforward, visual approach to comprehending intricate risk scenarios within a system. Its name originates from the distinctive shape of the diagram utilized, resembling a bowtie.

This methodology comprises several core components, including threats (or causes), barriers, top events, escalation factors, escalation factor barriers, and consequences. The "top event" in the BowTie model denotes the pivotal moment when control of the system is compromised, potentially resulting in hazardous situations.

On the left side of the bowtie (the left knot), threats and preventive barriers are delineated. These preventive barriers encompass controls implemented to forestall the occurrence of the top event. Conversely, on the right side (the right knot), potential consequences of the top event are outlined alongside recovery measures, which represent controls aimed at mitigating the consequences should the top event occur.

One of the advantages of the BowTie method lies in its visually intuitive layout, facilitating clear communication of risk scenarios across all organizational levels. It aids in identifying all conceivable threats that could precipitate a critical event, as well as the controls to preempt or alleviate these potential hazards.

However, while effective for high-level risk analysis and communication, the BowTie method may not explore the technical details as extensively as other risk assessment methodologies. It's often used in conjunction with more detailed risk assessment tools for a comprehensive risk evaluation.

The BowTie method can benefit also from a prioritized approach to barriers or control selection. The prioritized approach offers a strategic plan for security and compliance activities tailored to risk levels associated with ICS assets. This approach aims to assist organizations in creating a structured roadmap to address risks sequentially. It focuses on achieving "quick wins" through milestone target controls, facilitates financial and operational planning for compliance, ensures measurable progress in compliance initiatives, and fosters consistency within the organization.

According to *NIST 800-53, Revision 3*: *"This recommended sequencing prioritization helps ensure that foundational security controls upon which other controls depend are implemented first, thus enabling organizations to deploy controls in a more structured and timely manner in accordance with available resources."*

This approach is well defined, as depicted in *Figure 6.7*:

Priority Code	Sequencing	Action Priority Code
Priority Code 1 (**P1**)	FIRST	Implement P1 security controls first.
Priority Code 2 (**P2**)	NEXT	Implement P2 security controls after implementation of P1 controls.
Priority Code 3 (**P3**)	LAST	Implement P3 security controls after implementation of P1 and P2 controls.
Unspecified Priority Code (**P0**)	NONE	Security control not selected for baseline.

Figure 6.7 – Prioritized approach

For instance, if we combine BowTie and the prioritized approach, this might result in the following barriers and sequencing:

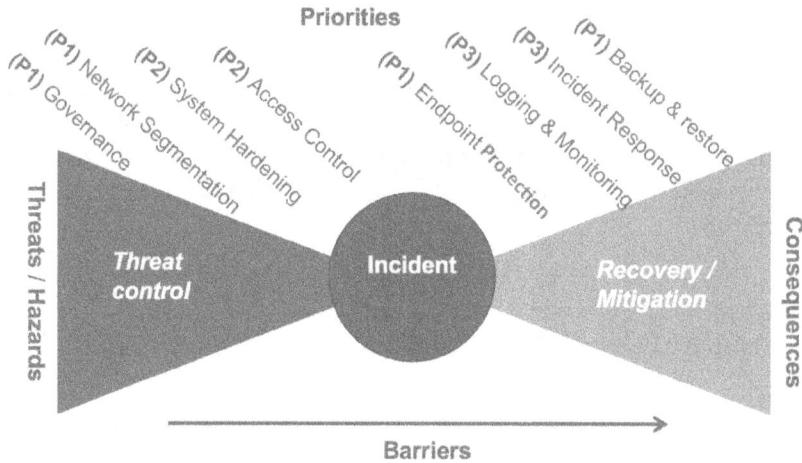

Figure 6.8 – Example of barriers with prioritized approach

It's clear that BowTie, with its visual and engineering-based approach, serves as a powerful tool to identify and highlight areas of risk within a system that require immediate attention, enabling engineers and risk managers to prioritize and address the most significant vulnerabilities effectively.

NAMUR

NAMUR is an international association of automation technology in process industries, and it provides several recommendations (*NE*) and worksheets (*NA*) aimed at the security risk assessment of SISs.

The NAMUR recommendations serve as general guidance to holistically consider security aspects in functional safety management. They emphasize the idea that safety and security should not be treated as separate entities but as complementary aspects to design and operate SISs.

In terms of carrying out a security risk assessment for an SIS, NAMUR provides a structure to systematically identify potential security risks and evaluate their impact on the safety functions of the SIS. Though it does not prescribe a specific method, it suggests that the security risk assessment should follow certain broad steps. These steps include the following:

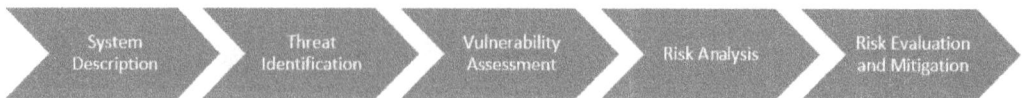

Figure 6.9 – NAMUR SIS cybersecurity risk assessment process

The NAMUR SIS cybersecurity risk assessment process includes the following aspects:

1. **System description**: The process begins by establishing a clear understanding of the target system. It includes the identification of all safety critical components and an understanding of the system architecture and its interaction with the external environment.

2. **Threat identification**: Next, all possible threats to the system are identified. These could stem from accidental incidents, malicious attacks, natural disasters, or system failures.

3. **Vulnerability assessment**: By examining each component of the system, it's possible to find vulnerabilities that could potentially be exploited by these threats. This process includes technical, procedural, and organizational vulnerabilities.

4. **Risk analysis**: Coming to a quantification stage, every identified vulnerability is evaluated based on the threats it faces and adverse impacts that could arise if these threats were actualized. This involves determining the likelihood of a threat exploiting a vulnerability and the resulting impact, leading to a risk rating.

5. **Risk evaluation and mitigation**: Using the risk rating, each risk is compared against predefined criteria to decide whether it's acceptable or if mitigation action is required. If risks are not found at an acceptable level, mitigative measures are identified and implemented to reduce the risk to an acceptable minimum.

The cybersecurity risk assessment could be executed using the *NA 163* checklist as a supportive tool. The most recent version of the checklist can be found on the NAMUR website: `https://www.namur.net/en/recommendations-and-worksheets/current-nena.html`.

The continuous nature of risk assessment

Risk assessment in the context of SISs and ICSs is not a one-off activity but a continuous, cyclic process. This is primarily due to the dynamic nature of both the technological environment and the threat landscape. System configurations, technological developments, employee behavior, and external threat factors are all variables that can change over time, affecting the risk profile of the system.

Regular updates to risk assessments are, therefore, crucial to maintaining an accurate picture of the security posture of the SIS or ICS. Changing network configurations, implementing new software or hardware, identifying new vulnerabilities, or the emergence of new types of cyber threats are all reasons to update a risk assessment. Further, maintenance activities, system upgrades, and significant changes in operating conditions or organizational structure may also necessitate a review. Regular reviews and updates help to ensure that the risk mitigation strategies remain effective and that appropriate response plans are in place.

The frequency of these reviews can vary depending on the specific circumstances of an organization, the criticality of the system, and the rate at which changes are happening. However, as a standard practice, at least an annual review is recommended. More frequent reviews could be necessary in a highly dynamic operational environment. It's important to note that risk assessment should not be seen only as a compliance exercise but rather as a critical component of a proactive and effective cybersecurity management strategy for SISs and ICSs.

Summary

This chapter provided a comprehensive overview of risk assessment in the context of SIS cybersecurity, elucidating the fundamental principles, various methodologies, and practical execution of such assessments. It underscored the importance of identifying SuCs, detecting threats and vulnerabilities, determining the possible consequences and impacts, assessing inherent risk, and establishing SL-Ts.

This chapter also revealed the relevance and potential applications of risk assessment methodologies such as NAMUR, *IEC 62443*, BowTie, and *NIST 800-82* in evaluating and mitigating SIS cybersecurity risks. It closely evaluated the iterative and dynamic process of effectively conducting risk assessments, highlighting the necessity for regular updates to stay responsive to emerging threats and changes in the operational environment.

Furthermore, it laid emphasis on the continuous nature of risk management, the need for re-evaluation of risks, and the requirement of ongoing attention to maintain robust and effective cybersecurity in SIS environments. The essence of this chapter signified the importance of integrating a proactive, structured, and continual risk assessment into the overall SIS cybersecurity strategy to facilitate a resilient and secure operational environment.

In the next chapter, we will explore the legislation and regulation landscape as well as emerging standards related to ICS cybersecurity and, in particular, SISs.

Further reading

- Demonstrating the safety of SISs: `https://www.cisa.gov/sites/default/files/2023-04/principles_approaches_for_security-by-design-default_508_0.pdf`

- *UK Government Security – Secure by Design Principles*: `https://www.security.gov.uk/guidance/secure-by-design/principles/`

- *Safety Instrumented Systems Engineering Handbook*: `www.kenexis.com_resources_TechnicalPapers_SIS_Handbook.pdf`

- *Defending critical infrastructure: The challenge of securing industrial control systems*: `https://cncpic.mai.gov.ro/sites/default/files/2022-09/20220602-Hybrid-CoE-Working-Paper-18-Defending-critical-infrastructure-WEB.pdf`

- Wireless instrumentation for SCSs:

  - `sintef-a26762-wireless-instrumentation-for-safety-critical-systems.-technology-standards-solutions-and-future-trends(1).pdf`

  - `us-18-Carcano-TRITON-How-It-Disrupted-Safety-Systems-And-Changed-The-Threat-Landscape-Of-Industrial-Control-Systems-Forever-wp.pdf`

  - `https://i.blackhat.com/us-18/Wed-August-8/us-18-Carcano-TRITON-How-It-Disrupted-Safety-Systems-And-Changed-The-Threat-Landscape-Of-Industrial-Control-Systems-Forever-wp.pdf`

- *Triton – The Deadly New Industrial Cyberweapon*: `triton-deadly-new-industrial-cyberweapon.pdf`

- *Guide to Industrial Control Systems (ICS) Security*: `https://csrc.nist.gov/pubs/sp/800/82/r2/final`

- *Current NAMUR Recommendations (NE) and Worksheets (NA)*: `https://www.namur.net/en/recommendations-and-worksheets/current-nena.html`

7
Security Standards and Certification

In the previous chapter, we journeyed through the world of Industrial Control Systems (**ICSs**) cybersecurity and have come to appreciate just how crucial security is – for everything from the design phase to the safe operations of mission critical systems, particularly, Safety Instrumented Systems (**SISs**).

As we delve deeper into the current era of cybersecurity, it is very promising to see that there is growing recognition of the crucial role of industrial cybersecurity from suppliers, end users, and even regulatory authorities.

Today, we find ourselves in a world where land, sea, and even space are now potential targets. This startling reality is further emphasized by the uplift in incidents that have been specifically aimed at industrial environments. Without a doubt, the industrial sector and **Critical Infrastructures** (**CIs**) have now become prime targets for cyber attacks.

Because of this, the insurance industry is beginning to rethink how it approaches its contracts concerning cybersecurity. With the risk of cyber attacks on industrial systems growing at a rapid pace, many insurance companies have responded by introducing specialized cyber insurance policies.

Levels of fear, meanwhile, are growing among the general public. As individuals, we are increasingly recognizing the risks posed to us from an online safety, security, and privacy perspective – and these concerns have caused the wheels of regulatory bodies to turn faster and faster. Regulators have been galvanized into action by the onslaught and severity of cyber threats in recent years – and we have seen several new acts and directives introduced and current ones revised and enforced, all to enhance our industrial cybersecurity and ensure greater compliance.

At a broader level, we are also witnessing the dawn of new regulatory requirements, designed to strengthen the compliance process. This trend has given a big boost to existing national and sector-specific regulations, pushing them toward adopting a more all-encompassing approach to ICS cybersecurity. The importance of being compliant with safety and cybersecurity regulations is now more evident than ever before, and an essential facet of the future of ICS.

Given this landscape, it's not hard to see why certification and assurance schemes are picking up steam, fast. They stand as important benchmarks to ensure standardization and consistency and to provide a minimum security baseline to fulfill the varied needs of everyone involved, including the regulators, end users, and suppliers. This growing focus reinforces how vital cybersecurity and compliance have become to the future of our ICS.

This chapter aims to examine the present landscape of regulations and legislation enveloping ICS cybersecurity, alongside the relevant certification schemes that are arising within this domain. We will delve into the different facets, the inherent challenges, and the opportunities that lie in fortifying the cybersecurity of these installations, with particular attention paid to the potential impact on SISs.

In this chapter, we will cover the following topics:

- The evolution of standards and legislation
- Industry-relevant certifications
- Identifying key stakeholders and the broader ecosystem
- Resources and initiatives

The evolution of standards and legislation

ICS cybersecurity has thrived in recent years. The domain has made significant strides forward when it comes to improving the maturity of the market and ensuring that the majority of these assets are deemed and classified as CIs in many countries. While the legislation and its level of enforcement vary widely forms depending on the regions and sectors, in this chapter, we will focus on a few examples that aim to cover all regions.

The legislations are usually mandated by competent authorities at a national level and are imposed in the form of acts, regulations, or directives. These standards are industry-specific and usually initiated by professional associations or working groups and might also consist of local or international members. Some of the standards are considered state-of-the-art or good engineering practices:

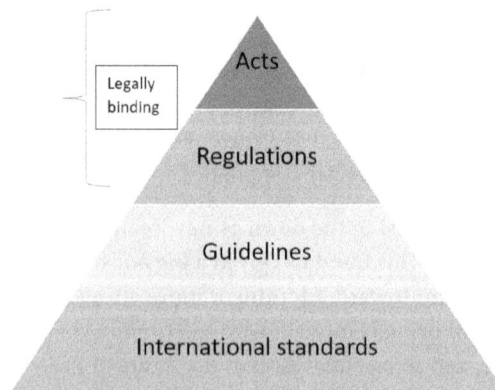

Figure 7.1 – Regulatory hierarchy concerning ICS cybersecurity

Let's take a look at the regulatory hierarchy that's relevant to cybersecurity in ICS:

- **Acts**: In the legislative context, an act is a statutory law that's passed by a legislative body, either at the local or national level. In terms of cybersecurity and safety, an act usually provides the legal basis for regulations and enforcement actions to protect information and infrastructure. It can lay out penalties for non-compliance and can apply not only to corporations but also to individuals as well.

  An example would be the **Executive Order on Improving the Nation's Cybersecurity 14028** in the United States, which has the objective of enhancing cybersecurity through a variety of initiatives related to the security and integrity of the software supply chain, and the **Cyber Resilience Act (CRA)** in the EU, which addresses the cybersecurity of hardware and software products (including SIS) with a focus on manufacturers, distributors, and importers.

- **Regulations and directives**: Both of these are legislative instruments that aim to enforce certain acts or laws. Regulations are binding legislative acts that have to be applied in their entirety across all relevant jurisdictions. They are often more specific than acts and are used to implement the objectives of the acts, including specifics about how compliance can be achieved and verified. In the context of cybersecurity, an example would be the **General Data Protection Regulation (GDPR)** of the European Union. Directives, on the other hand, also bind countries to the goals to be achieved but give them freedom in determining how to achieve these objectives. They must be incorporated into the national laws of each country. The following are some examples:

  - **Network and Information Security (NIS) EU Directive 2**: The NIS2 is an EU directive that aims to boost the overall level of cybersecurity across the European Union. It is an update to the original NIS Directive of 2016 and continues the work of protecting CIs in the EU.

  - Introduced in 2020, the NIS2 broadens the scope of its predecessor to cover more sectors. In addition to the original sectors, which include energy, transport, and healthcare, NIS2 also covers new sectors, such as social networks, cloud computing services, and some public administration services. The directive not only imposes strict cybersecurity requirements to these sectors but also sets forth obligations for these entities to report significant cyber incidents to their respective authorities. It also enforces stronger supervisory measures and incident response requirements for these services on a national and pan-EU level.

  - Furthermore, NIS2 introduces stricter penalty systems for non-compliance and encourages greater cooperation between EU states. This comes in the form of sharing of vital information about cyber threats, vulnerabilities, incidents, and solutions, thus promoting a culture of collaboration and mutual assistance. In summary, the NIS2 Directive is a major step forward by the EU to raise its cyber-resilience, as well as foster a unified approach to tackle cybersecurity challenges across all member states.

- **CRA**: Currently proposed by the European Union, the CRA is an ambitious legislation that aims to strengthen the overall digital resilience of the EU's critical entities and networks. The act mainly intends to improve the resilience and response capabilities of essential entities in case of cyber threats and attacks. It plans to achieve this by setting clear responsibilities for both public and private stakeholders, driving cooperation and information sharing between them, and ensuring they have the necessary operational capabilities. The CRA intends to bring a substantial number of sectors under its umbrella, including energy, transport, banking, financial market infrastructures, healthcare, drinking water, wastewater, digital infrastructure, and public administration, among others. Moreover, the CRA proposes streamlining the reporting obligations of these sectors to shape a more uniform and less burdensome system and foster an effective information flow on cross-border incidents that engages all relevant actors. In a nutshell, the EU's CRA aims to create a robust framework that fortifies the defenses of essential entities against cyber threats, thus ensuring a safer and more resilient digital environment across the European Union.

- **Guidelines**: Guidelines provide recommendations and general directions to help individuals or organizations achieve compliance with acts, regulations, or standards. They are usually non-binding and serve to provide best practices, frameworks, or processes for entities to follow to ensure compliance or to achieve certain benchmarks. For example, the **Cybersecurity Framework** from the US's **National Institute of Standards and Technology** (**NIST**) is a set of guidelines that organizations can use to manage and reduce their cybersecurity risk.

- **International standards**: These are generally agreed-upon frameworks, practices, or specifications that are defined by international standardization bodies such as the International **Organization for Standardization** (**ISO**) or the **International Electrotechnical Commission** (**IECN**). International standards such as the ISA/IEC 62433 series for industrial cybersecurity provide globally recognized principles and practices for establishing, maintaining, and improving **Cybersecurity Management Systems** (**CSMSs**), as well as for the design and architecture of suppliers' cybersecurity. These standards can play a critical role in helping businesses achieve regulatory compliance and inspire confidence among customers, partners, and stakeholders.

Industrial cybersecurity and safety are regulated by a wide variety of legislations, standards, and guidelines around the world. Here's a list of key regulations and compliance requirements that pertain to this field:

- **IEC 61508**: The IEC 61508 standard, issued by the **International Electrotechnical Commission** (**IEC**), presents a consolidated framework for the functional safety of **Electrical/Electronic/ Programmable Electronic** (**E/E/PE**) safety-related systems. This standard, although primarily focused on safety, implicitly addresses cybersecurity as a contributing factor toward achieving overall system safety. It emphasizes **Security Risk Assessment** (**SRA**) during the safety lifecycle, encompassing activities such as hazard and risk analysis, design, implementation, and the operation and maintenance of safety systems. IEC 61508 seeks to mitigate safety risks, including those posed by cyber threats that could potentially lead to dangerous failures or malfunctions in safety critical applications. Furthermore, it underscores the need to consider the security aspects during the design and implementation of safety systems to increase the resilience and robustness of these systems against cyber threats. The application of this standard ultimately supports both the secure operation of industrial environments and the preservation of functional safety amid increasing cybersecurity risks.

- **IEC 61511**: The IEC 61511 standard, issued by the IEC, provides guidelines for the application of SIS within the process sector. Although primarily focused on functional safety, IEC 61511 recognizes the interdependencies that exist between safety and security, particularly within the context of cyber threats. The standard introduces the concept of a safety lifecycle, which provides a structured approach to safety that includes activities such as risk analysis, design, implementation, and the operation and maintenance of SIS. Within this lifecycle, security risk assessment is emphasized, during which all security aspects are fully taken into account to ensure that safety functions are not compromised. IEC 61511 encourages the application of a **Defense-in-Depth** (**DiD**) security approach, reinforcing how crucial it is to implement multiple layers of protection to defend against cyber threats. This holistic approach helps to enhance the resilience of SIS against cyber threats, thereby reducing the risk of systems failure or malfunctions that can have dangerous consequences. Ultimately, by considering both safety and security aspects, the IEC 61511 standard supports the secure and safe operation of process industries, reinforcing the need for built-in resilience against escalating cyber risks.

- **API 1164**: The 3rd edition of the API Standard 1164 establishes guidelines for Pipeline Control Systems Cybersecurity, released by the **American Petroleum Institute** (**API**). The key focus of this standard is to manage and mitigate cyber threat risks inherent in Industrial **Automation and Control** (**IAC**) settings. The ultimate objective is to enhance security, fortify system integrity, and uphold resiliency. It aims to safeguard pipeline infrastructures by heightening the digital and operational safety measures in place to bolster safety and avert possible disruptions in supply chains. The recent, expanded update now also incorporates new cybersecurity prerequisites and elevated risk assessment protocols.

- **IEC TS 63074**: This technical specification is an authoritative document issued by the IEC. This specification dedicates itself to the intersection of functional safety and cybersecurity aspects, specifically within the realm of machine design. Its primary objective is to effectively manage cyber-related risks that could potentially compromise the functional safety of machinery by mitigating identified risks. Among its key features is to provision detailed guidelines for machinery manufacturers, which offer a standardized procedure to identify and manage security-related risks during the initial design stages of machinery. IEC TS 63074 extends its reach to cover the machinery's entire lifecycle, from its conceptual design and development phase through to its decommissioning and disposal phase. The standard's comprehensive approach enhances overall system security, safeguards functional safety, and minimizes disruptions to operational processes.

- **North American Electric Reliability Corporation Critical Infrastructure Protection (NERC CIP) (USA and Canada)**: Enforceable standards for the bulk electric system in North America.

- **Network and Information Systems (NIS) Directive (EU)**: The first piece of EU-wide legislation on cybersecurity, it provides legal measures to boost the overall level of cybersecurity in the EU.

- **Regulation on the Security of Network and Information Systems (UK)**: National law based on the NIS Directive.

- **National Institute of Standards and Technology (NIST) (USA and Globally)**: This set of voluntary guidelines helps organizations manage and reduce cybersecurity risk to CI.

- **The International Society of Automation (ISA) (Globally)**: The ISA/IEC 62443 series of standards provide a framework to address and mitigate security vulnerabilities in **Industrial Automation and Control Systems (IACSs)**.

- **Australian Signals Directorate's Essential Eight (Australia)**: A set of strategies to help mitigate cybersecurity incidents.

- **The Industrial Internet Consortium (IIC) (Globally)**: Provides several frameworks and reference architectures including the **Industrial Internet Security Framework (IISF)**.

Please note that this is not an exhaustive list. The regulations that an organization must legally comply with may vary based on the regions of operation and the specific industry they belong to.

Industry-relevant certifications

Today, the IEC 62443 series is the most relevant and globally adopted industrial cybersecurity standard in the industry. IEC 62443 is a series of standards developed by the IEC for IACS security. With growing interconnectivity and the rapid advancement and evolution of technology, ICS are increasingly vulnerable to cybersecurity threats. The IEC 62443 standards provide a comprehensive approach to improve the security of these systems across various areas such as risk assessment, system design, procedures, and personnel competencies.

The standards are divided into four main parts: General, Policies and Procedures, System, and Component. Each part consists of several standards that are designed to address specific facets of ICS security. The IEC 62443 series provides a common language and systematic methodology for users, suppliers, and integrators to maximize the security capabilities of ICS.

The following figure outlines the relevant IEC 62443 within associated ecosystems, as well as the key certifications:

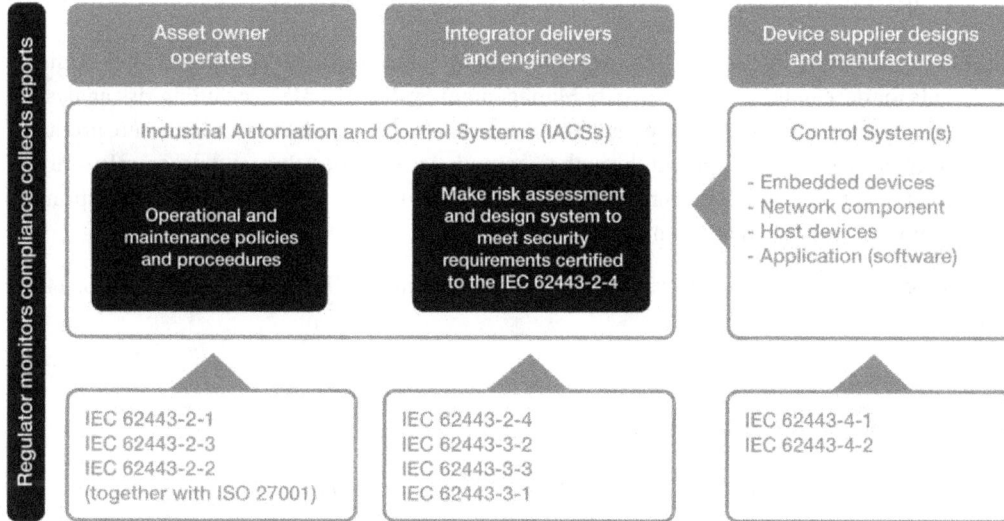

Figure 7.2 – Structure of the IEC 62443 standards

An important feature of the IEC 62443 is that it provides a **Security Level (SL)** system that classifies facilities into four levels (SL 1 to SL 4) based on the types of threats they may encounter. This helps organizations to understand their risk profile and identify appropriate security measures:

- **Asset owners**: For the asset owners and operators, the benefits of the IEC 62443 certification are substantial. The standards guide them to strategically identify and manage risks associated with their ICS. By following the standards, end users can ensure that their systems are designed, installed, maintained, and operated in a way that appropriately manages cybersecurity risk. Compliance with these standards also mandates clear requirements for third parties and suppliers to mitigate supply chain attacks and partially transfer the risk to these parties while increasing the reliability of their ICS operations.

- **Suppliers and manufacturers**: For suppliers and system integrators, the IEC 62443 certification enhances the trustworthiness of their products in the marketplace as well as their security practices, including processes. It assures potential customers about the high-quality security features of their products. Supplier's products, when designed to meet these strict standards, are likely to be much more resilient to cyber threats and more compatible with systems that also adhere to these standards. Furthermore, through certification, suppliers can differentiate their products and stand out in an increasingly competitive market.

- **System integrators**: They can ensure that the quality of services provided is in line with expectations and meets industry quality standards – demonstrating well-defined processes and qualified personnel that are capable of delivering secure solutions.

IEC 62443 standards provide a solid framework to develop robust security capabilities. They outline requirements for developing a **Cybersecurity Management System (CSMS)**, including risk analysis, DiD strategies, and security incident response. The standards also have guidelines for the secure product lifecycle, from design and development through to commissioning, operating, and decommissioning systems. This systematic and comprehensive approach enhances security readiness and response capabilities both at the system and component levels.

At the time of writing, ISASecure provides a certification scheme for IEC 62443-4-1 and IEC 62443-4-1 that includes the following aspects:

- **Component Security Assurance (CSA)**

- **System Security Assurance (SSA)**

- **Secure Development Lifecycle Assurance (SDLA)**

The following links provide an overview of the certification scheme and a list of certified SIS components:

- *End Users*: https://isasecure.org/end-users

- ISA/IEC 62443-4-2 Certified Components: https://isasecure.org/end-users/iec-62443-4-2-certified-components

Within the IEC 62443 series of standards, the SLs define four increasing levels of security assurance that a product or system can attain. These assurance levels are measured on a scale of 1 to 4 – called SL1 to SL4, respectively – where each level corresponds to an increasing security assurance requirement. Here is a brief explanation of each level:

SL	Description
SL 1	Designed to protect against casual or coincidental violations mainly because of a lack of awareness by people who have the authority to perform an action. It covers protection against basic cyber threats and is applicable where the risk/exposure is relatively small.
SL 2	Intended to address intentional violation using simple tools and methods by people who have the authority to perform an action. It covers protection against well-known threats and applies where moderate/high-risk exposure exists but still within the low sophistication of attackers.

SL	Description
SL 3	Protects against intentional violation using sophisticated tools and methods by individuals or groups who do not necessarily have the authority to perform an action but have low skill levels. The threat scenario includes advanced threats and is applicable where the risk/exposure is significant, requiring protection against skilled attackers.
SL 4	Addresses deliberate violations using sophisticated tools and methods by individuals or groups who have the resources to perform an action and are highly skilled. This level pertains to high-risk scenarios such as nation-state attacks. It covers protection against the most advanced threats and is considered for very high-risk and high-security-critical applications.

Table 7.1 – SLs

It's worth mentioning that the level of security required depends on the individual risk assessments by organizations, always considering the potential impact of a security incident. Higher SLs require more extensive safeguards and a higher degree of assurance than the lower levels. However, reaching higher SLs also demands more resources and costs while possibly affecting system usability and performance. Therefore, determining the appropriate SAL is a balance that must be struck between risk management and operational impact.

Moreover, the IEC 62443-4-1 and IEC 62443-4-2 standards specifically provide security requirements for the product development lifecycle and technical security requirements for IACS components, respectively. Being validated by a third party for these certifications can provide a competitive edge by demonstrating a provider's commitment to secured product development and supply chain management.

The IEC 62443 certification is not merely an asset but an essential marker of trusted functionality and assurance for both end users and suppliers in an era where cybersecurity threats are ever evolving. Much more than simply adhering to a set of practices, it signifies a commitment to continuously improving security posture and maintaining an effective ICS that can withstand emerging cyber threats.

The IEC 62443 series of standards provides comprehensive guidance on the security aspects of IACS. Specifically, IEC 62443-4-1, IEC 62443-4-2, and IEC 62443-2-4 address critical aspects of product development, technical security requirements, and a supplier's development lifecycle and practices, respectively:

- **IEC 62443-4-1**: The **Secure Product Development Lifecycle Requirements** standard provides process requirements for the secure development of products used in an IACS and defines a secure development lifecycle for the design and development of new or updates to existing products. This includes aspects such as defining security requirements, secure design, secure implementation (including coding guidelines), verification and validation, defect management, patch management, and product end-of-life.

- **IEC 62443-4-2**: The **Technical Security Requirements for IACS Components** standard provides guidance on the detailed technical security requirements for components that comprise ICS, including embedded devices, network components, host components, and software applications. The focus is on the capabilities that components should have to resist cyber threats, grouped by their required SLs.

- **IEC 62443-2-4**: The **Security Program Requirements for IACS Service Providers** standard sets the requirements for establishing and maintaining a cybersecurity management system for suppliers providing assets, software, and services for IACS. The aim is to ensure that suppliers securely develop and deliver their products, and then provide ongoing support securely.

Each of these specifications serves as a crucial link in the chain of ensuring robust ICS cybersecurity. Compliance with these standards assures stakeholders that necessary security controls are being appropriately implemented across all stages, from the product development phase to the assurance of robust technical security requirements, and throughout the suppliers' lifecycle.

For more information about the certification of systems, especially SIS, please visit the ISASecure website: `https://isasecure.org/certification`.

While acknowledging the guidance that these existing standards and certifications offer, it is essential to also remember that the landscape of industrial cybersecurity is constantly changing. The highly dynamic nature of cyber threats, the onward march of technology, and the increasing interconnectivity of industrial systems necessitate a robust, adaptive standards framework that never stops evolving.

Recognizing the importance of adapting and conforming to these emerging standards and certifications is vital as we move toward a new era of ICS. Not only do they provide the required assurance of cybersecurity but they also help guide the industry forward when it comes to navigating the many challenges and complexities of securing ICS.

Prioritizing adherence to these standards and understanding the certifications will pave the way for the safety, efficiency, and future resilience of ICS environments.

Identifying key stakeholders and the broader ecosystem

Industrial cybersecurity involves various key stakeholders, who each play their unique roles and interact with one another to ensure the secure operation of ICS. Here's a breakdown of these key stakeholders and how they are all interconnected:

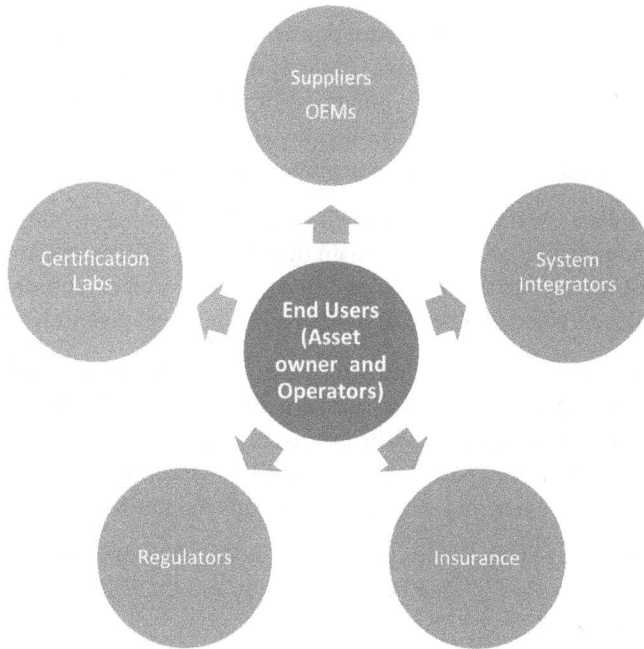

Figure 7.3: Key stakeholders

The following list outlines the ecosystem of stakeholders and their primary responsibilities:

- **End users**: This group primarily includes companies and organizations that run ICS. They are responsible for identifying their cybersecurity needs, implementing necessary measures, complying with regulations, and providing necessary training to their staff. They directly interact with suppliers, OEMs, and regulators while also considering the inputs from insurance companies.

- **Operators**: Also, part of the end user group, operators are the individuals who interact with the ICS on a day-to-day basis. Their role is crucial in following cybersecurity protocols, identifying and reporting potential issues, and implementing operational aspects of cybersecurity practices.

- **Suppliers**: Suppliers provide the hardware, software, services integration, and additional necessary components that make up the ICS. They must ensure that their products meet the required safety and security standards and comply with relevant regulations. Suppliers often work in tandem with both OEMs in the manufacturing process and with end users to meet their specific requirements.

- **Original equipment manufacturers (OEMs)**: OEMs design and produce the machinery that is used in industrial processes. They are responsible for integrating supplied components (often inclusive of ICS) securely, following cybersecurity best practices. OEMs closely work with suppliers in the design and manufacturing process to provide tailored equipment to the end users.

- **Insurance companies**: Specialist insurance policies are now offered that cover losses related to cyber threats. So, insurance companies evaluate the cybersecurity readiness of end users' systems, suggest risk reduction measures, and provide policies for associated incident coverage. Insurance companies also communicate directly with regulators to better understand legal requirements, all of which influence their insurance offerings.

- **Certification bodies**: Certification bodies play a crucial role in promoting standardization and enforcing compliance within the domain of ICS. These institutions are responsible for providing independent validation and certification that verifies an organization's, product's, or individual's compliance with defined standards, regulations, and protocols.

    In the context of ICS, certification bodies focus on evaluating the design, implementation, and operation of systems against a set of established criteria, such as the IEC 62443, IEC 61508/IEC 61511, and ISO/IEC 27001 standards. The process may involve examining the cybersecurity management systems, auditing operational processes, inspecting the control environment, reviewing personnel qualifications, and conducting vulnerability and risk assessments.

    Some prominent bodies that provide certifications in the realm of ICS cybersecurity include entities such as DNV, Exida, TÜV Rheinland Group, and the **Underwriters Laboratories (UL)**, to name just a few.

- **Regulators**: Regulators now play a pivotal role in the industry in terms of setting the rules, standards, and regulations that must be adhered to by all stakeholders. They set out the **minimum security requirements** that end users, suppliers, and OEMs must all meet to be compliant. They also directly communicate with insurance companies to help assess policies' compliance with the law.

The interactions among these stakeholders form a strong network whereby each player's decisions and activities can greatly influence others. For example, when regulators update standards, all other stakeholders must adjust their strategies and practices to stay compliant. An insurance company's assessment can lead to an end user having to increase their cybersecurity measures, which, in turn, impacts their requirements from suppliers and OEMs.

This interactive network of stakeholders ensures a holistic approach is taken toward industrial cybersecurity, where all aspects from design and manufacturing to operation and coverage in case of breaches are taken into account.

> **Certifications benefits**
>
> The certification process provides several benefits. For end users and operators, certification by a recognized body provides confidence in the security measures implemented by their suppliers. For suppliers and manufacturers, being certified can give them a competitive edge by demonstrating their compliance with recognized security standards and best practices, while for insurance companies and investors, it facilitates risk assessment and decision-making.
>
> Lastly, for regulators and government bodies, it can assist in monitoring and enforcing compliance and standards within the ICS landscape.
>
> It is important to always bear in mind that obtaining certification is not a one-time process; regular audits and reviews are required to ensure continuous compliance and improvement within the rapidly evolving ICS cybersecurity landscape.

Resources and initiatives

The vast digital world offers an abundance of reputable resources that cater specifically to professionals working with ICS, **Supervisory Control and Data Acquisition (SCADA)**, and SIS. These encompass professional associations, government bodies, international organizations, and academic institutions that extend their expertise to help strengthen cybersecurity in these critical systems.

In this section, we have collated some of these resources and will guide you toward trusted sources of information that can provide further detail on best practices, strategies, and accepted standards prevalent in the field of SIS cybersecurity:

- The **International Society of Automation (ISA)** serves as a leading global authority, providing its members with resources such as the ISA/IEC 62443 series of standards developed explicitly for ICS cybersecurity. The ISA also offers a host of training and certification programs for professionals in the field: `https://www.isa.org/standards-and-publications/isa-publications`.

- **NAMUR**, an international association of automation technology users in process industries, provides a platform to exchange ideas and best practices. They also offer practical recommendations, worksheets, and reports relevant to SIS security: `https://www.namur.net/en/index.html`.

- The **WIB Process Automation Users' Association** provides process instrumentation evaluation and assessment services for international vendors and end users, enhancing the overall security and reliability of these systems: `https://www.wib.nl/home`.

- The Electric Power Research Institute (EPRI) carries out **research, development, and demonstration (RD&D)** activities for the electricity sector worldwide, addressing the cybersecurity challenges associated with energy generation, distribution, and grid systems: `https://www.epri.com/research`.

- The **Idaho National Laboratory** (**INL**), part of the US Department of Energy's complex system of research laboratories, has extensive resources dedicated to ICS cybersecurity, including research papers, guides, and training initiatives: `https://inl.gov/research-programs/`.

- **Linking the Oil and Gas Industry to Improve Cybersecurity** (**LOGIIC**) is an ongoing collaboration of oil and natural gas companies and the US Department of Homeland Security, Science and Technology Directorate. LOGIIC's primary mission is to facilitate collaborative R&D among cybersecurity vendors, focusing on the protection of ICS and related systems: `https://logiic.org/projects`.

- **Cybersecurity and Infrastructure Security Agency** (**CISA**) provides a wide range of resources, including alerts and advisories, best practices, recommended standards, and training, targeting key stakeholders in the ICS community to protect CI: `https://www.cisa.gov/topics/industrial-control-systems`.

- The **UK Health and Safety Executive** (**HSE**) and the UK **National Cyber Security Centre** (**NCSC**) provide guidelines for industrial safety and cybersecurity, with regular updates to reflect changing threat landscapes:

  - `https://www.hse.gov.uk`

  - `https://www.ncsc.gov.uk`

- The **Netherlands' National Cyber Security Center** (**NL NCSC**) provides a wealth of information and resources targeting various aspects of cybersecurity, including specific resources for ICS protection: `https://english.ncsc.nl`.

- The **SANS Institute** offers a range of resources for ICS cybersecurity, along with a repository of research papers and webinars authored by industry experts in this domain: `https://www.sans.org/security-resources/?msc=main-nav`.

- **ISAsecure** is a globally recognized certification program that assures cybersecurity for ICS products and systems, ensuring that they are robust against network attacks and free from known vulnerabilities: `https://isasecure.org/learning-center`.

- **Sandia National Laboratories and Pacific Northwest National Laboratory** (**PNNL**) offer extensive research programs that cover a broad range of areas related to energy, national security, and environmental science, including cybersecurity:

  - `https://www.sandia.gov/resilience/res-testbed/`.

  - `https://www.pnnl.gov/cybersecurity`.

- The **European Union Agency for Cybersecurity** (**ENISA**) fosters a safer internet by providing recommendations on cybersecurity, cooperating with member states, and building capacity within the EU: `https://www.enisa.europa.eu/`.

- The **European Energy – Information Sharing & Analysis Centre** (**EE-ISAC**) is a major resource for organizations in the energy sector that provides threat intelligence feeds, sharing best practices, and supporting workshops and webinars on various cybersecurity topics: `https://www.ee-isac.eu`.

Connecting with these varied resources will significantly enrich and update your ICS cybersecurity knowledge. These trustworthy sources provide insightful guidelines, regular updates on emerging threats, necessary tools, and training opportunities that will enhance your capabilities when it comes to managing and ensuring SIS security.

In an ever-shifting cyber threat landscape, such tools, tips, and training are invaluable for tackling and neutralizing cybersecurity threats, thereby ensuring the continuous, efficient, and safe operation of the critical industrial systems that we all rely on.

Summary

The dynamism of emerging global directives and standards highlights the ongoing strides that are being made to secure the CIs that power our world. The growing convergence of safety and cybersecurity considerations in these standards underscores the importance that is now being given to adopting a holistic, integrated approach to tackling the big challenges at hand.

In this chapter, we explored the intersection of regulations, standards, and certifications and their impacts on the field of ICS cybersecurity. We also traced the evolution of standards and legislation, addressing the changes that have been driven both by technological advances and the ever-maturing threat landscape.

A closer look at industry-relevant certifications has underscored their significance in assuring compliance and endorsing competence while analyzing certification bodies demonstrated the indispensable role that these bodies play not just in validating but also in inspiring trust in ICS cybersecurity initiatives. Amid the rising cybersecurity concerns that exist within the ICS landscape, such certifications have become much more than just a compliance requirement – they serve as a testimony of an organization's commitment to safety and security.

This chapter also explored the interconnection of key stakeholders in the ICS cybersecurity arena, from end users, suppliers, and operators to insurance companies and regulatory bodies. We established that a proactive collaboration among these entities is of paramount importance; only by sharing our learnings can we collectively build a resilient ICS ecosystem that is capable of facing, and overcoming, the cybersecurity challenges of the 21st century.

As we conclude this chapter, the importance of ongoing vigilance is clear. Only through the adoption of the latest standards and certifications, and a unified collaboration among all stakeholders, can we ensure the safety, security, and resilience of ICS. We are now entering a new era in which all parties need to adjust and adapt to a brand-new wind of change on the horizon. In the following chapter, we will delve into the innovative initiatives and future developments in ICS and SIS cybersecurity.

Further reading

To learn more about the topics that were covered in this chapter, take a look at the following resources:

- The European CRA: `https://www.european-cyber-resilience-act.com/`

- CSA Singapore Cybersecurity Act: `https://www.csa.gov.sg/legislation/Cybersecurity-Act`

- *EXECUTIVE ORDER 14028, Improving the nation's cybersecurity*, `https://www.nist.gov/itl/executive-order-14028-improving-nations-cybersecurity`

- *API Standard 1164, 3rd Edition*: `https://www.api.org/products-and-services/standards/important-standards-announcements/1164`

- *Shifting the Balance of Cybersecurity Risk: Principles and Approaches for Security-by-Design and -Default*: `https://www.cisa.gov/sites/default/files/2023-04/principles_approaches_for_security-by-design-default_508_0.pdf`

- Secure by Design Principles: `https://www.security.gov.uk/guidance/secure-by-design/principles/`

- ISASecure Certifications: `https://isasecure.org/certification`

- IEC TS 63074:2023: Safety of machinery – security aspects related to the functional safety of safety-related control systems: `https://webstore.iec.ch/publication/69228`

- Executive Order on Improving the Nation's Cybersecurity: `https://www.whitehouse.gov/briefing-room/presidential-actions/2021/05/12/executive-order-on-improving-the-nations-cybersecurity/`

The Future of ICS and SIS: Innovations and Challenges

In an era that has been dominated by rapid technological evolution and widespread digitalization, **Industrial Control Systems (ICSs)** and **Safety Instrumented Systems (SISs)** are no exceptions. We are currently in a period of vast transformation globally – and with this comes some profound challenges. Digital innovation is playing a powerful role in revolutionizing numerous lives worldwide, while at the same time making us susceptive to some previously unforeseen risks, at both an individual and organizational level.

This delicate balance between progress and vulnerability underscored by digitalization has existed for a while now. However, its influence on the function of critical infrastructures and other safety critical systems is heightening at a breathtaking pace. Alongside, ICS – responsible for overseeing and managing production facilities – are battling an increasing array of threats fortified by increasingly persistent malicious actors including national entities.

Within the context of such a volatile landscape, it is vital that we are proactively aware of the many concerns and factors that are now compelling organizations toward implementing countermeasures that help to ensure the safety, security, and longevity of their ICS domains.

This final chapter offers a glimpse into the many challenges and innovations that these systems are set to encounter in the near future, including some of the current patterns within the ICS security sector that are chiefly focused on technology. In addition, this chapter clarifies the existing scenarios that exist within the ICS security sphere, as well as providing some discerning observations on the potential ICS security tendencies that it is predicted could emerge in the very near future.

This heightened insight provides a window into the future of the ICS security realm, enlightening us on the paths we must take and the hurdles we might encounter along the way when navigating this challenging landscape, with knowledge and preparedness on our side.

In this chapter, we will focus on the following topics:

- The current state of ICS cybersecurity innovation
- Emerging technologies including opportunities and challenges

The current state of ICS cybersecurity innovation

The industrial sector, characterized by its static and conservative nature, continues to move forward at a sluggish pace when it comes to technological innovation and adoption.

This environment is heavily influenced by numerous long-established practices and a focus on maintaining the status quo with current, familiar technologies. Consequently, there is a noticeable reluctance to introduce novel products onto the plant floor, particularly within the realm of **Safety Instrumented Systems (SISs)**. The domain of SIS is particularly constrained, dominated by a limited number of players and burdened by extensive certification and testing processes.

These requirements often stifle innovation, creating a significant disparity between the potential pace of technological progress and the actual advancements within the industry. This gap hinders the sector from fully realizing the benefits and opportunities that modern technology could bring to truly enhance business operations for the better.

Currently, many industrial environments continue to depend on outdated security technologies such as firewalls, anomaly detection systems, and application whitelisting. These methods, once considered cutting-edge, are now failing to adequately protect mission critical operations against more sophisticated threats. Meanwhile, emerging technologies such as artificial intelligence, advanced quantum cryptography, and Zero-Trust architectures remain largely inaccessible. These innovations are widely perceived as distant and impractical options due to the inherent challenges they present, including the need to overhaul obsolete and vulnerable existing systems – a process that requires substantial investment and a strategic operational shift.

Moreover, **Original Equipment Manufacturers (OEMs)** and suppliers in this arena are often not the first to embrace new technologies. This hesitance is partly due to their comfortable market positions and the lack of competitive pressure they experience; this is especially true for SIS vendors. Furthermore, many of the new technologies being proposed originate from IT-centric backgrounds, which may not seamlessly integrate into ICS environments. This mismatch can lead to a number of implementation challenges that further discourage adoption in a sector where reliability and stability are of paramount importance.

Over the last decade, there has been a significant focus on network logging and monitoring to increase visibility and enhance asset inventories. This area has matured substantially within industrial cybersecurity, with a proliferation of products from companies including Nozomi, Claroty, SecurityMatters, and Dragos.

These technologies have been integrated into ICS suppliers' solutions, marking one of the few areas of innovation within this field. However, the market is still dominated by these incumbents and radical, game-changing technologies are scarce. Recent years have also seen a rapid consolidation in this space, with major OEMs such as Honeywell, Rockwell, and Siemens acquiring smaller, innovative products to bolster their cybersecurity offerings.

Despite these advancements, the broader market for disruptive cybersecurity technologies remains limited; the integration challenges and the high cost of entry for new technologies continue to be significant barriers. These issues are compounded by the fact that many ICS environments operate with legacy systems that are not only difficult to upgrade but are also often incompatible with modern cybersecurity solutions. This incompatibility underscores the critical need for sector-specific solutions that address the unique requirements of industrial environments.

In conclusion, the industrial cybersecurity landscape is currently caught in a cycle of cautious technology adoption, with a heavy reliance on outdated methods and the slow integration of advanced technologies. This conservative approach is partly due to the unique challenges that exist within the sector, including the need for rigorous testing and certification, the risk of obsolescence, and the operational implications of deploying new technologies.

To bridge this gap and foster a more dynamic approach to cybersecurity, the industry must not only invest in newer technologies but also cultivate a much more competitive and innovative ecosystem that encourages technological advancement and better secures critical infrastructures. To achieve this, industry leaders must prioritize the development of an integrated cybersecurity strategy that addresses both IT and **Operational Technology (OT)** aspects. This strategy should include the deployment of advanced cybersecurity technologies that are specifically designed for the unique challenges of the ICS environment.

Additionally, there must be an increased focus on cybersecurity education and training for all stakeholders involved in industrial operations – from engineers to IT staff – to foster a more aware and proactive cybersecurity culture. Partnerships between industry, academia, and government also have a pivotal role to play when it comes to advancing cybersecurity innovations. These collaborations can help to standardize security protocols, share critical threat intelligence, and develop new technologies that are tailored to the specific needs of the industrial sector. By working together, these entities can accelerate the adoption of advanced cybersecurity.

In the following section, we will delve into the emerging technologies poised to revolutionize ICS. We'll examine how these innovations can empower asset owners and operators, enhance business operations, and provide competitive advantages for OEMs and suppliers. This exploration will highlight the transformative potential of these technologies in reshaping the industrial landscape.

Emerging technologies including opportunities and challenges

Over the past few years, the industry landscape has undergone a huge transformation due, in large part, to the rapid and widespread adoption of new technologies. This shift has been largely driven by several global factors: the COVID-19 pandemic, which necessitated new ways to work and interact; the energy transition, which seeks to move away from fossil fuels towards more sustainable energy sources; and the digital transformation, which is revolutionizing how businesses operate across all sectors.

These technologies are not only introducing innovative use cases but are also enhancing operational efficiency, boosting productivity, and providing solutions to critical challenges such as the shortage of skilled and qualified personnel.

However, the introduction of these technologies has also presented a double-edged sword to both organizations and society at large. While the latest technological advancements offer the vast potential to bolster both defensive and offensive capabilities, they simultaneously pose significant security risks. This paradox has led to what can be described as a race condition – a scenario where rapid technology adoption can lead to security vulnerabilities if not managed properly. Both nation-states and various actors in cyberspace are now in a constant battle to fully understand and master these technologies to better leverage their capabilities while also mitigating potential threats.

To aid the wider industry, we have compiled a list of the key technology innovations that will be shaping industrial environments – and will likely be adopted by the majority of asset owners, operators, OEMs, and suppliers – in the near future.

Artificial Intelligence

Artificial Intelligence (**AI**) is rapidly gaining traction as a pivotal tool in securing ICS. Integrating AI into ICS security protocols can significantly improve both the efficacy and efficiency of these protective measures. Nonetheless, the application of AI in this context is not without its risks and challenges.

The potential to optimize industrial processes is remarkable, but the uncertainty and unpredictability associated with AI decisions could complicate safety considerations and result in unforeseen vulnerabilities.

For instance, GenAI is being adopted by the majority of industries. AI copilots are being productized into every piece of OEM technology. AI has given rise to the term "hallucination," which means false or misleading data being presented by the AI as fact. This can cause major issues if users are leveraging these copilots to make decisions.

We explore the potential impact of AI on ICS and SIS cybersecurity in the following table, covering the benefits and risks related to its use:

Benefits	Challenges
Improving safety and reliability: AI algorithms proactively monitor and identify potential safety hazards well before they become actual risks to operations or personnel, highlighting AI's critical role in advancing safety protocols within modern industrial settings.	**Qualified workforce**: The complex nature of AI systems demands a workforce proficient in the design, deployment, and maintenance of these sophisticated technologies, necessitating specialized training and expertise.
Real-time decision making: AI's capacity to quickly process and interpret data allows it to make well-informed decisions that refine production operations and address potential setbacks promptly. This capability not only boosts operational efficiency but also strengthens the system's resilience against unexpected events.	**High-quality data**: Effective AI implementation relies heavily on the availability of high-quality data. Variabilities in data quality, especially in ICS, can compromise the precision of AI-driven outputs.
Proactive threat identification: AI's ability to sift through extensive data sets enables it to spot unusual patterns and behaviors indicative of security threats, thus facilitating faster and more accurate responses from security teams.	**Vulnerability to adversarial attacks**: AI systems are susceptible to adversarial attacks where malicious inputs are designed to confuse and bypass AI, leading to erroneous outcomes such as false positives or negatives, which can impair the AI's effectiveness.
Network security: AI tools are instrumental in analyzing network behavior to detect and thwart suspicious or unauthorized activities, thereby safeguarding ICS from potential cyber intrusions that threaten critical infrastructure.	**Lack of transparency**: AI solutions often lack clear interpretability in their operations, which can complicate the trust and validation processes for security professionals relying on AI outputs.
Predictive maintenance efficiency: By analyzing operational data, AI can predict the need for maintenance and component replacements in advance, helping to avert machine failures and associated operational downtime or safety incidents.	**Legal issues**: AI can introduce legal and regulatory compliance, privacy, and copyright issues if not deployed properly.
Automation of security processes: AI can automate mundane and repetitive security tasks such as log analysis and vulnerability assessments, freeing up security personnel to concentrate on more strategic issues such as threat hunting and incident management.	**False positives**: While AI enhances threat detection, it can also lead to false positives—alerts about non-existent threats, which can cause "alert fatigue" and potentially divert attention from genuine threats.

Benefits	Challenges
Enhancing data privacy measures: The integration of AI helps bolster data privacy protocols by ensuring systematic and secure data handling, critical in maintaining confidentiality and integrity within ICS environments.	The complexity of data storage and processing is an issue, along with the unknown long-term consequences of potential issues.

Table 8.1 – AI benefits and challenges

AI holds transformative potential for ICS security, offering advancements in threat detection, predictive maintenance, network security, and the automation of routine security tasks. Nonetheless, deploying AI in ICS security is not without obstacles; challenges such as false positives, opacity in decision-making processes, inconsistent data quality, and susceptibility to adversarial attacks must be addressed. Security professionals must judiciously assess both the advantages and drawbacks of integrating AI into their operations.

To optimize AI's efficacy and ensure its secure application, adopting best practices is crucial. These practices include the careful selection of appropriate AI algorithms, the guarantee of high-quality data inputs, AI governance and the integration of human oversight, and continuous testing of AI systems to affirm their effectiveness and uncover any vulnerabilities. By striking a balance between these benefits and challenges, security teams can enhance their protective measures for critical infrastructure and bolster their response to emergent security threats.

Quantum computing

Quantum computing represents a seismic shift in the landscape of computational technology, with the potential to deliver computational power far surpassing the capabilities of today's most advanced supercomputers. This leap in processing power is not merely incremental; it is a fundamental transformation that promises to solve complex problems within seconds—problems that currently take even the fastest conventional computers years to solve. For ICS, which are pivotal to the infrastructure of utilities such as power plants, water treatment facilities, and transportation systems, the implications of quantum computing are profound, both in terms of the potential benefits and the significant challenges it poses.

The potential benefits of quantum computing in ICS cybersecurity include the following:

- **Enhanced security protocols**: Quantum computers have the potential to run complex algorithms that can strengthen cryptographic measures far beyond the capabilities of classical computers. For example, **Quantum Key Distribution (QKD)** offers a theoretically unbreakable encryption method, which could fundamentally secure communication channels between critical infrastructure components.

- **Superior threat detection**: The ability of quantum computers to process large datasets rapidly allows for real-time threat analysis and detection at speeds unachievable with current technology. This capability could be used to identify potential threats and vulnerabilities in ICS much quicker, enabling preemptive action against cyber-attacks.

- **Simulation and optimization of network operations**: Quantum computing could optimize the operations of ICS through more efficient logistics and system management, potentially reducing costs and increasing reliability. Algorithms designed for quantum computers could model complex networks and predict outcomes with high accuracy, leading to optimized energy usage and distribution logistics.

- **Tokenization and encryption**: In the context of ICS, tokenization can anonymize direct references to sensitive data, thereby enhancing security. Quantum computing could further enhance this process through more complex and secure tokenization algorithms that are difficult for conventional computers to decode. Meanwhile, quantum encryption could provide a layer of security that is virtually impenetrable by today's standards, utilizing the principles of quantum mechanics to encrypt data in a way that any attempt to eavesdrop would automatically alter the state of the data, thus revealing the eavesdropper.

On the other hand, there are challenges and risks associated with quantum computing in ICS that we should be aware of:

- **Threat to current cryptography**: Quantum computing poses a grave risk to the cryptographic frameworks that currently protect global financial markets, confidential data, and ICS communications. Quantum computers have the potential to break widely used cryptographic algorithms such as RSA and ECC, which would compromise the integrity and confidentiality of sensitive information within ICS.

- **Quantum-safe cryptography development**: To counteract the threats posed by quantum capabilities, there is a pressing need for quantum-resistant or quantum-safe cryptography. Developing these new cryptographic systems is complex and resource intensive. They must not only secure systems against quantum attacks but also be efficient enough to be implemented with the current technological infrastructure without degrading system performance.

- **Implementation costs and complexity**: The integration of quantum computing technologies into existing ICS architectures poses significant logistical and financial challenges. The cost of quantum computers and the complexity of quantum information systems are currently prohibitive for widespread adoption. Additionally, integrating these systems into existing cybersecurity frameworks requires substantial modification to existing protocols and systems, which can be a costly and complex process.

It's worth mentioning that technology adoption in ICS, particularly SIS, progresses at a notably sluggish pace due to its critical nature and the emphasis on safety. Only systems that have been rigorously tested and proven effective in similar environments are permitted for deployment. Consequently, the process of integrating new technologies into SIS often spans several years.

As we inch closer to the quantum age, the strategic importance of quantum computing for cybersecurity in critical infrastructures such as ICS cannot be overstated. The transition to quantum-safe cryptography needs to be a priority for cybersecurity professionals working within ICS domains. This involves not only upgrading cryptographic measures but also ensuring that all elements of the cybersecurity infrastructure are robust against quantum threats.

Moreover, the integration of quantum computing into ICS will require a collaborative effort among tech developers, cybersecurity experts, and industrial operators to ensure that the deployment of quantum technologies is seamless and secure. Education and training will be paramount in preparing the workforce for the upcoming quantum revolution in industrial cybersecurity.

In conclusion, while the advent of quantum computing brings with it formidable challenges, particularly in the realm of cryptography, the potential benefits it offers in terms of security enhancement, threat detection, and system optimization are immense. The race to quantum readiness is not just about thwarting threats but also about harnessing quantum technology to fortify and advance the capabilities of ICS. The journey towards a quantum-secure industrial landscape is complex and fraught with challenges, but with proactive preparation and strategic investment in quantum-safe technologies, it is a feasible goal that can secure the future of industrial cybersecurity.

Cloud computing

Cloud computing is increasingly being seen as a transformative technology for ICS and SIS, offering substantial benefits such as scalability, cost efficiency, and enhanced data management capabilities. These advantages allow organizations to dynamically scale their infrastructure according to demand without incurring significant upfront costs, optimize operational processes, and improve overall efficiency. However, the migration of ICS and SIS to cloud environments also introduces several big cybersecurity challenges that must be carefully managed.

Some of the benefits of cloud computing in ICS and SIS cybersecurity include the following:

- **Enhanced scalability and flexibility**: Cloud platforms provide ICS and SIS with the ability to scale resources on-demand, adapting to changes in processing needs without the need for significant capital investment. This flexibility supports the dynamic nature of industrial operations, allowing for the efficient management of resources during varying levels of demand.

- **Cost-effectiveness**: By leveraging cloud computing, organizations can reduce costs associated with maintaining and upgrading physical servers and other infrastructure. The pay-as-you-go model of cloud services means costs are directly related to consumption, providing a more economical approach to data management and processing.

- **Improved data analytics and insights**: Cloud environments enhance the capabilities of ICS and SIS in terms of data collection and analytics. With advanced cloud-based analytics tools, companies can process vast amounts of data to glean insights that help in predictive maintenance, operational efficiency, and decision-making processes.

- **Disaster recovery and data redundancy**: Cloud services offer robust disaster recovery solutions and data redundancy. This is crucial for ICS and SIS environments where data integrity and availability are critical for operational continuity and safety.

In the meantime, cloud computing also introduces some challenges in ICS domains, such as the following:

- **Data privacy and security concerns**: As data moves to the cloud, ensuring its privacy and security becomes more challenging. The distributed nature of cloud computing can dilute the control organizations have over their own data. Sensitive information handled by ICS and SIS, if compromised, can lead to significant safety and financial risks.

- **Loss of control over data**: Handing over infrastructure management to third-party cloud service providers can lead to a perceived loss of control. This is particularly concerning for industries that operate under strict regulatory compliance regarding data handling and processing.

- **Potential for increased cyber threats**: Cloud platforms can be targets for cyberattacks due to the valuable data they store. Ensuring that cloud deployments adhere to stringent cybersecurity standards is essential to protect against potential breaches, especially given the critical nature of the systems involved.

- **Integration issues**: Integrating legacy ICS and SIS with modern cloud solutions can be technically challenging. Compatibility issues may arise, requiring additional resources or modifications to existing systems, potentially disrupting operations.

- **Compliance and regulatory challenges**: Adhering to industry-specific regulations can be more complicated with cloud-based systems. Compliance audits can become more frequent and demanding, requiring organizations to demonstrate that they handle data securely and in accordance with all applicable laws and standards.

In summary, the impact of cloud computing on ICS and SIS cybersecurity is both profound and multifaceted. While the benefits can significantly enhance operational efficiency and scalability, the many challenges necessitate a careful and strategic approach to cloud adoption.

Organizations must develop comprehensive cloud security strategies that encompass robust encryption practices, regular security assessments, continuous compliance checks, and effective incident response plans. Additionally, partnering with cloud service providers that understand the unique needs of industrial systems and offer tailored security measures can mitigate many of the risks that are closely associated with cloud computing.

Furthermore, cloud computing is not suitable for SIS due to concerns related to reliability, availability, latency, security, regulatory compliance, control, visibility, and organizational factors.

As cloud technology continues to evolve, so too must the cybersecurity strategies employed by industries relying on ICS and SIS. By proactively addressing these challenges and capitalizing on the opportunities provided by the cloud, organizations can secure their critical infrastructure while achieving greater operational efficiency and resilience.

Autonomous operations

Autonomous operations in ICS and SIS represent a significant shift towards more efficient and reliable industrial processes. By leveraging automation, these systems can operate with greater precision and minimal human intervention, leading to enhanced productivity and reduced operational costs. The integration of autonomous technologies allows for real-time data processing and decision-making, optimizing performance and predictive maintenance, which are crucial for critical infrastructures such as manufacturing plants, energy grids, and water treatment facilities.

Let's explore some of the benefits of autonomous operations in ICS environments:

- **Increased operational efficiency**: Automated systems streamline complex processes, reduce human error, and ensure operations are carried out swiftly and efficiently. This not only improves overall productivity but also enhances system reliability and process safety, which are paramount in industrial settings.

- **Enhanced predictive maintenance**: Autonomous systems can predict failures before they occur by analyzing data patterns and performance metrics. This predictive capability ensures timely maintenance actions, prevents unexpected downtimes, and extends the lifespan of critical equipment.

- **Scalability and flexibility**: Automation allows for easier scalability and adaptability of systems. As operational demands change, autonomous systems can adjust more fluidly than manual setups, making them ideal for industries experiencing rapid growth or those needing to adjust operations in response to market or environmental changes.

Of course, there are challenges related to the deployment of autonomous operations:

- **Complexity of cyber threat vectors**: With autonomy comes complexity, especially in terms of the cybersecurity threat landscape. Autonomous systems introduce new vectors for cyber-attacks, as they often require integration with cloud computing platforms and IoT devices, broadening the attack surface significantly.

- **Liability and accountability issues**: Determining liability in the event of a failure or security breach becomes more complicated with autonomous systems. The diffusion of responsibility between system providers, operators, and third-party service providers can lead to significant legal and regulatory challenges, particularly when trying to pinpoint the source of a malfunction or breach.

- **Dependence on data integrity**: The effectiveness of autonomous operations is heavily dependent on the quality and integrity of data. Any manipulation or corruption of data, whether through cyber-attacks or technical faults, can lead to incorrect decisions by the automation algorithms, potentially causing severe operational disruptions or safety hazards.

- **Resistance to change and integration difficulties**: Implementing autonomous systems often requires significant changes in existing processes and infrastructures. There can be resistance to such changes from within the organization due to the fear of job displacement or mistrust of fully automated systems. Additionally, integrating new technologies with legacy systems presents technical and operational challenges.

The potential impact of autonomous operations on ICS and SIS cybersecurity is both transformative and complex. While the benefits of improved efficiency, reliability, and predictive maintenance are clear, these must be balanced against the challenges of increased cyber threat vectors, liability issues, and data dependency. As the industry moves towards greater autonomy, it is critical to develop robust cybersecurity frameworks that can adapt to and mitigate the risks associated with these advanced technologies. This includes implementing advanced threat detection systems, ensuring data integrity, and developing clear protocols for liability and accountability. By addressing these challenges head-on, industries can fully leverage the benefits of autonomous operations while securing their critical infrastructures against potential threats.

Zero Trust

The Zero Trust security model, which operates on the principle that no entity inside or outside the network should be automatically trusted, is increasingly being recognized as a vital framework for enhancing cybersecurity in ICS and SIS. This paradigm shift towards a "never trust, always verify" approach is particularly relevant for industries with critical infrastructures that face a high risk of sophisticated cyber threats.

Let's examine Zero Trust adoption in ICS and SIS cybersecurity:

- **Enhanced security posture**: Zero Trust significantly tightens security by rigorously authenticating and continuously validating every request to connect to the system, regardless of origin. This minimizes the attack surface and reduces the likelihood of unauthorized access.

- **Micro-segmentation**: This core component of Zero Trust involves dividing security perimeters into small zones to maintain separate access for separate parts of the network. If a system is compromised, micro-segmentation limits the potential lateral movement of attackers within the network, thereby protecting critical control system components.

- **Least privilege**: The least-privilege principle in Zero Trust ensures that users and systems are granted only the minimum access necessary to perform their functions, enhancing security by reducing potential attack surfaces and making systems such as ICS and SIS more resilient to intrusion and attack.

- **Improved compliance**: Adopting a Zero Trust architecture can aid in compliance with stringent regulatory requirements, as it provides a structured approach to data security and privacy that aligns with global cybersecurity standards.

The following are some of the challenges of Zero Trust implementation in ICS environments:

- **Integration complexity**: Implementing Zero Trust in environments traditionally governed by legacy systems and proprietary protocols presents significant challenges. The architecture demands extensive modifications to existing networks, which can be costly and complex, and might disrupt operational processes.

- **Cultural and operational shifts**: Zero Trust requires a fundamental change in how security is approached; it necessitates a cultural shift within the organization towards continuous monitoring and strict access controls. This can be difficult to achieve, particularly in sectors where there is heavy reliance on trusted legacy systems and processes.

- **Continuous monitoring and maintenance**: Zero Trust architectures depend on real-time threat detection and continuous monitoring, which require sophisticated tools and can incur substantial ongoing operational costs. Ensuring these systems are always up to date and functioning can be a logistical challenge.

The adoption of a Zero Trust model in ICS and SIS environments represents a proactive step toward defending critical infrastructures against increasingly sophisticated cyber threats. While the transition involves substantial challenges, including integration complexities and the need for cultural shifts within organizations, the potential benefits of improved security, compliance, and breach containment are compelling. For industries reliant on ICSs and SISs, moving towards Zero Trust not only enhances operational security but also builds a robust foundation for future technological integrations. As such, organizations should begin by assessing their current security measures, consulting with cybersecurity experts, and planning strategic, phased implementations to minimize disruptions while maximizing the effectiveness of Zero Trust measures.

Self-healing systems

Self-healing systems represent a revolutionary advancement in the cybersecurity landscape for ICS and SIS, offering the potential to significantly bolster system resilience. These systems are designed to automatically detect, diagnose, and respond to disruptions, including cyber-attacks, effectively minimizing downtime and mitigating potential damage. This capability not only enhances operational continuity but also improves the overall security posture by reducing the window of opportunity for attackers.

Self-healing systems can positively impact ICS and SIS cybersecurity by providing the following:

- **Enhanced system resilience**: By integrating self-healing mechanisms, ICS and SIS can rapidly recover from attacks without human intervention, thereby maintaining critical operations under adverse conditions. This resilience is crucial for industries where system downtime can lead to significant economic losses or safety hazards.

- **Proactive threat management**: Self-healing systems proactively manage threats by continuously monitoring for anomalies and automatically implementing corrective actions. This ongoing vigilance helps prevent the escalation of potential security breaches, keeping systems secure in a dynamic threat environment.

Some challenges associated with self-healing systems are as follows:

- **Loss of control**: One of the primary concerns with self-healing systems is the potential loss of control over critical processes. As these systems take autonomous actions, operators may feel detached from the decision-making process, leading to challenges in oversight and governance.

- **Misinterpretation of benign actions**: There is also the risk that self-healing systems might misinterpret benign actions as malicious, leading to unnecessary or incorrect responses. This could disrupt normal operations and result in operational inefficiencies or, worse, unintended safety incidents.

The integration of self-healing technologies into ICS and SIS environments presents a promising frontier for enhancing cybersecurity resilience. However, to fully capitalize on their benefits while mitigating associated risks, it is crucial to establish robust parameters and control mechanisms that ensure these systems operate within defined operational norms and contribute positively to overall system security. Moreover, continuous refinement of the algorithms that govern self-healing actions will be essential to minimize errors in threat perception and response. In conclusion, while self-healing systems offer substantial advantages in terms of resilience and proactive security, they require careful implementation and ongoing management to truly align with the critical safety and operational needs of ICS and SIS environments.

Finally, the future landscape of ICS and SIS cybersecurity is poised to undergo a transformative integration of cutting-edge technologies aimed at enhancing defense mechanisms across all operational levels. This advanced cybersecurity architecture will necessitate a defense-in-depth strategy that not only layers multiple protective barriers but also deeply embeds security within every stratum, from enterprise systems to direct operational controls. Secure integration of IT and ICS environments will be crucial in ensuring comprehensive coverage against cyber threats. This IT/OT convergence must be handled with an acute awareness of the distinct challenges and risks presented by these interfaced domains. The architecture must not only be robust in its defensive capabilities but also remain agile and adaptive – continually evolving in sync with new technological advancements and shifting threat landscapes.

By harnessing these advanced technologies – AI, ML, quantum computing, Zero Trust architectures, autonomous operations, and self-healing systems – the next generation of ICS and SIS cybersecurity can achieve unprecedented levels of security and efficiency. These systems will not only defend against current cyber threats but also anticipate and neutralize emerging risks, thereby securing critical industrial infrastructures in an increasingly volatile cyber domain.

Summary

In this chapter, we delve into the current state of technological innovation within ICS and examine the disruptive technologies poised to redefine the future of cybersecurity while offering significant business advantages. We have uncovered how these innovations not only promise a competitive edge but also introduce a spectrum of risks and challenges, some of which remain undefined at this stage. Through this exploration, we aim to provide a comprehensive understanding of both the opportunities and the potential pitfalls that accompany the integration of these advanced technologies into the ICS landscape.

Further reading

- *NIST Trustworthy & Responsible AI Resource Center*: https://airc.nist.gov/home

- **Cybersecurity and Infrastructure Security Agency (CISA)**: https://www.cisa.gov/ai

- **Centre for Internet Security (CIS)**: https://www.cisecurity.org/insights

- *Global Future Council on the Future of Artificial Intelligence*: https://www.weforum.org/communities/global-future-council-on-artificial-intelligence/

- *Evaluating Cryptographic Vulnerabilities Created by Quantum Computing in Industrial Control Systems*: https://www.rand.org/pubs/research_reports/RRA2427-1.html

- *Moving Toward an All-of-the-Above Approach to Quantum Cybersecurity*: https://www.csis.org/analysis/moving-toward-all-above-approach-quantum-cybersecurity

- *Quantum Computing and Cybersecurity*: https://www.belfercenter.org/publication/quantum-computing-and-cybersecurity

- *Post-Quantum Cryptography Initiative*: https://www.cisa.gov/quantum

- *Global Future Council on the Future of Quantum Economy*: https://www.weforum.org/communities/gfc-on-quantum-economy/

- *Self-Healing in Cyber–Physical Systems Using Machine Learning: A Critical Analysis of Theories and Tools*: https://www.mdpi.com/1999-5903/15/7/244

- *Adaptive Immunity for Software: Towards Autonomous Self-healing Systems*: https://www.researchgate.net/publication/351501922_Adaptive_Immunity_for_Software_Towards_Autonomous_Self-healing_Systems

- *SASH: Safe Autonomous Self-Healing*: https://link.springer.com/chapter/10.1007/978-3-031-26507-5_12

- *An Autonomous Self-learning and Self-adversarial Training Neural Architecture for Intelligent and Resilient Cyber Security Systems*, https://link.springer.com/chapter/10.1007/978-3-031-34204-2_38

Index

‹packt›

packtpub.com

Subscribe to our online digital library for full access to over 7,000 books and videos, as well as industry leading tools to help you plan your personal development and advance your career. For more information, please visit our website.

Why subscribe?

- Spend less time learning and more time coding with practical eBooks and Videos from over 4,000 industry professionals

- Improve your learning with Skill Plans built especially for you

- Get a free eBook or video every month

- Fully searchable for easy access to vital information

- Copy and paste, print, and bookmark content

Did you know that Packt offers eBook versions of every book published, with PDF and ePub files available? You can upgrade to the eBook version at packtpub.com and as a print book customer, you are entitled to a discount on the eBook copy. Get in touch with us at customercare@packtpub.com for more details.

At www.packtpub.com, you can also read a collection of free technical articles, sign up for a range of free newsletters, and receive exclusive discounts and offers on Packt books and eBooks.

Other Books You May Enjoy

If you enjoyed this book, you may be interested in these other books by Packt:

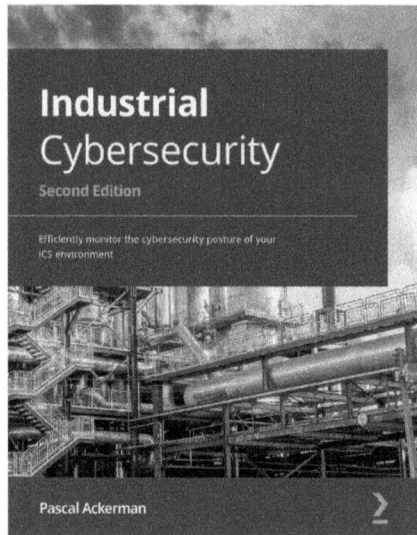

Industrial Cybersecurity

Pascal Ackerman

ISBN: 978-1-80020-209-2

- Monitor the ICS security posture actively as well as passively
- Respond to incidents in a controlled and standard way
- Understand what incident response activities are required in your ICS environment
- Perform threat-hunting exercises using the Elasticsearch, Logstash, and Kibana (ELK) stack
- Assess the overall effectiveness of your ICS cybersecurity program
- Discover tools, techniques, methodologies, and activities to perform risk assessments for your ICS environment

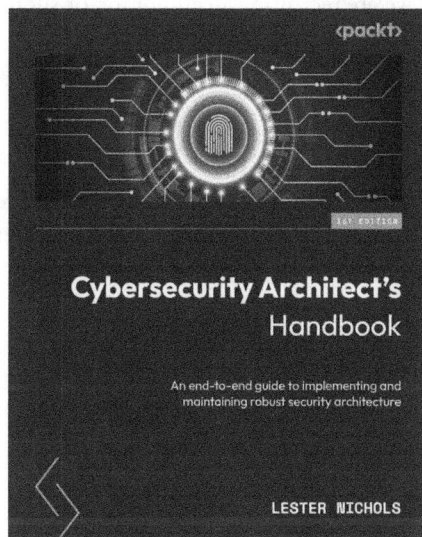

Cybersecurity Architect's Handbook

Lester Nichols

ISBN: 978-1-80323-584-4

- Get to grips with the foundational concepts and basics of cybersecurity
- Understand cybersecurity architecture principles through scenario-based examples
- Navigate the certification landscape and understand key considerations for getting certified
- Implement zero-trust authentication with practical examples and best practices
- Find out how to choose commercial and open source tools
- Address architecture challenges, focusing on mitigating threats and organizational governance

Packt is searching for authors like you

If you're interested in becoming an author for Packt, please visit authors.packtpub.com and apply today. We have worked with thousands of developers and tech professionals, just like you, to help them share their insight with the global tech community. You can make a general application, apply for a specific hot topic that we are recruiting an author for, or submit your own idea.

Share Your Thoughts

Now you've finished *Securing Industrial Control Systems and Safety Instrumented Systems*, we'd love to hear your thoughts! Scan the QR code below to go straight to the Amazon review page for this book and share your feedback or leave a review on the site that you purchased it from.

https://packt.link/r/1-801-07881-5

Your review is important to us and the tech community and will help us make sure we're delivering excellent quality content.

Download a free PDF copy of this book

Thanks for purchasing this book!

Do you like to read on the go but are unable to carry your print books everywhere?

Is your eBook purchase not compatible with the device of your choice?

Don't worry, now with every Packt book you get a DRM-free PDF version of that book at no cost.

Read anywhere, any place, on any device. Search, copy, and paste code from your favorite technical books directly into your application.

The perks don't stop there, you can get exclusive access to discounts, newsletters, and great free content in your inbox daily

Follow these simple steps to get the benefits:

1. Scan the QR code or visit the link below

https://packt.link/free-ebook/978-1-80107-881-8

2. Submit your proof of purchase
3. That's it! We'll send your free PDF and other benefits to your email directly

www.ingramcontent.com/pod-product-compliance
Lightning Source LLC
Chambersburg PA
CBHW081100220326
41598CB00038B/7165